Electronics II

Electronics II

A textbook covering the original
Level II syllabus of the
Technician Education Council

D C Green

M Tech, CEng, MIERE
Senior Lecturer in Telecommunication Engineering
Willesden College of Technology

Second Edition

Pitman

PITMAN BOOKS LIMITED
128 Long Acre, London WC2E 9AN

Associated Companies
Pitman Publishing Pty Ltd, Melbourne
Pitman Publishing New Zealand Ltd, Wellington

© D. C. Green 1982

First published in Great Britain 1978
Reprinted 1979, 1981, 1982
Second edition 1982
Reprinted 1983

Text set in 10/12 Linotron Times,
printed and bound in Great Britain
at The Pitman Press, Bath

ISBN 0 273 01827 2

Contents

Preface to the Second Edition

The new TEC schemes for Electronic and Telecommunication Technicians no longer include an Electronic unit at level II. Instead the fundamentals of semiconductor theory and transistor circuitry are included in a new $1\frac{1}{2}$ unit level II Principles. Basic digital techniques now appear in a new half unit also at level II.

It is anticipated, however, that many Colleges will continue to offer an Electronic unit at level II and may well choose to do so including some reference to the field-effect transistor. It is unlikely that the thermionic valve will receive much, if any, attention, and accordingly the second edition of this book has replaced the chapter on valves with one on field-effect transistors. Also, the few examples that were given of valve circuitry have been replaced by their fet equivalents. Otherwise the main content of the book has been left unchanged.

The original learning objectives for the Electronic II unit have been retained as a guide to the student but the valve sections have been deleted.

DCG

Preface

The study of electronics is an essential part of the education of the electrical technician since the applications of modern electronics are widespread and are continually increasing. In the telecommunication field the use of electronic circuitry in both radio and line communication systems is well established and increasingly telephony systems are making use of digital electronic techniques.

The Technician Education Council (TEC) has introduced a new scheme for the education of technicians employed in the Electronic and Telecommunication industries. The scheme consists of a number of standard units that a technician is expected to be able to complete within a period of 3 years part-time study.

The first introduction to electronics is made at the second level and the unit, known as Electronics II, introduces the student to the basic principles of electronics which should be well understood before progress to practical circuits and systems can be made.

This book has been written to provide a complete coverage of the topics included in the TEC second level subject Electronics II. Chapter 1 outlines the basic theory of semiconductors and Chapters 2, 3, 4 and 5 introduce the reader to, respectively, the semiconductor diode, the bipolar transistor, the thermionic valve, and the cathode ray tube. These are all electronic devices in common use and a knowledge of their operation is necessary before an understanding of electronic circuits can be gained. Chapters 6 and 7 then present the basic principles of operation of audio-frequency amplifiers and of oscillators. An acquaintance with digital techniques is increasingly important for all electronic/telecommunication technicians and a simple introduction to the subject is given in Chapter 8. Finally, no electronic equipment is able to operate without a suitable power supply and Chapter 9 describes the operation of the common basic circuits.

This book, therefore, provides a text that constitutes a comprehensive introduction to the basic principles of electronics and should be suitable for any first course in electronics for technicians.

The contents of the unit Electronics II have been written by the Technician Education Council in the form of learning objectives that specify the goals of the unit and the means by which the student ought to be able to demonstrate his achievement of each goal. The learning objectives for Electronics II are given in an appendix at the end of the book and acknowledgement is made to the TEC for their permission to use the content of the unit. The council reserve the right to amend the contents of its unit at any time.

Many worked examples are provided in the text to illustrate the principles that have been discussed and each chapter concludes with a number of exercises and short exercises. Each exercise is expected to take a student approximately 30 minutes to answer, while the short exercises should occupy a much shorter time. Many of these exercises have been taken from part City and Guild examination papers and grateful acknowledgement of permission to do so is made to the Institute. Answers to the numerical problems are to be found at the end of the book; these answers are the sole responsibility of the author and are not necessarily endorsed by the Institute.

DCG

1 Simplified Semiconductor Theory

A semiconductor is defined as a material whose resistivity is much less than the resistivity of an insulator yet is much greater than the resistivity of a conductor, *and* whose resistivity decreases with increase in temperature. For example, the resistivity of copper is 10^{-8} ohm-metre, of quartz is 10^{12} ohm-metres, and for the semiconductor materials of interest in this chapter, that of silicon is 0.5 ohm-metre and that of germanium is 2300 ohm-metres at 27°C. To gain an appreciation of the operation of semiconductors and semiconductor devices, it is necessary to have some familiarity with the basic concepts of the atomic structure of matter.

A Simple Outline of Atomic Theory

All the substances which occur in Nature consist of one or more basic elements; a substance containing more than one element is known as a compound. An ELEMENT is a substance that can neither be decomposed (broken into a number of other substances) by ordinary chemical action, nor made by a chemical union of a number of other substances. A compound consists of two or more different elements in combination and has properties different from the properties of its constituent parts. Water, for example, is a compound of oxygen and hydrogen. A MOLECULE is the smallest amount of a substance that can occur by itself and still retain the characteristic properties of that substance, and may consist, for example, of two atoms of hydrogen and two atoms of oxygen for hydrogen peroxide, of one atom of oxygen and one atom of carbon for carbon monoxide, and of two atoms of oxygen and one atom of carbon for carbon dioxide. An ATOM is the smallest unit of which a chemical element is built. The atoms of any particular element all have the same average mass and this average mass differs from the average mass of the atoms of any other element.

*E & PROTONS HAVE
EQUAL CHARGE
∴ 2 PROTONS CANCELS
THE EFFECTS of 2
ELECTRONS (E)*

The elements are grouped in an arrangement known as the Periodic Table of the Elements (Table 1.1) according to their chemical properties. Elements with similar properties are placed in the same vertical column. More than 100 elements are known to science today; some of them exist in large quantities and are commonly found throughout the world, e.g. oxygen, hydrogen and carbon, while others such as gold, uranium and radium are relatively rare, and some do not exist at all in Nature and are artificially created in equipment operated by atomic physicists.

An atom of any element consists of a complex pattern of electrons orbiting around a positively charged nucleus. The ELECTRONS each have a negative charge equal to 1.602×10^{-19} coulomb (known as the electronic charge e) and exist in just sufficient number to make the overall electrical charge of the atom equal to zero. The NUCLEUS itself consists of a certain number A of particles known as nucleons. A is the mass number of the atom. There are two kinds of nucleons: PROTONS, which each have a positive charge of e coulomb and NEUTRONS, which have zero charge. The number of protons in a nucleus is known as the atomic number of the atom, symbol Z, and the number of neutrons the neutron number N.

Hence $A = Z + N$.

The difference between the atoms of the various elements is in the number and arrangement of the electrons, protons and neutrons of which the atoms are composed. There is no difference between an electron in one element and another electron in any other element.

The Hydrogen Atom

The simplest atom is that of the element hydrogen and it consists merely of a single proton in the nucleus and a single electron in orbit around it (Fig. 1.1a). The helium atom is the next simplest and can be seen from Fig. 1.1b to consist of a nucleus containing two protons and two neutrons, with two electrons orbiting around it.

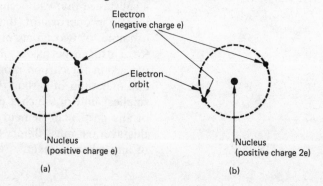

Fig. 1.1 The hydrogen and helium atoms

Electron
(negative charge e)

Electron
orbit

Nucleus
(positive charge e)

(a)

Nucleus
(positive charge 2e)

(b)

Table 1.1

I	II	III	IV	V	VI	VII	VIII	0
Hydrogen 1								Helium 2
Lithium 3	Beryllium 4	Boron 5	Carbon 6	Nitrogen 7	Oxygen 8	Fluorine 9		Neon 10
Sodium 11	Magnesium 12	Aluminium 13	Silicon 14	Phosphorus 15	Sulphur 16	Chlorine 17		Argon 18
Potassium 19	Calcium 20	Scandium 21	Titanium 22	Vanadium 23	Chromium 24	Manganese 25	Iron 26, Cobalt 27, Nickel 28	
Copper 29	Zinc 30	Gallium 31	Germanium 32	Arsenic 33	Selenium 34	Bromine 35		Krypton 36
Rubidium 37	Strontium 38	Yttrium 39	Zirconium 40	Niobium 41	Molybdenum 42	Technetium 43	Ruthenium 44, Rhodium 45, Palladium 46	
Silver 47	Cadmium 48	Indium 49	Tin 50	Antimony 51	Tellurium 52	Iodine 53		Xenon 54
Caesium 55	Barium 56	Rare Earths 57–71	Hafnium 72	Tantalum 73	Tungsten 74	Rhenium 75	Osmium 76, Iridium 77, Platinum 78	
Gold 79	Mercury 80	Thallium 81	Lead 82	Bismuth 83	Polonium 84	Astatine 85		Radon 86
Francium 87	Radium 88	Actinide Series 89–100						

For an electron to be able to move in a circular path around a nucleus, as shown in Fig. 1.1, it must have a force exerted on it pulling it towards the nucleus. This force is the electrical attractive force exerted by the positive nucleus on the negative electron. Work must be done in moving an electric charge through an electric field and so some work must have been done to move the electron from the nucleus to the orbit in which it is travelling. Thus an electron must possess a certain, discrete, quantity of energy before it can exist in an orbit around the nucleus. An electron can only exist in certain orbits of particular radii, and when in one of these orbits it must then possess the particular amount of energy associated with that orbit. An electron cannot occupy any orbit other than one of the orbits allowed as shown in Fig. 1.2. Normally the electron travels the innermost orbit since this is the orbit of least energy, but if it is given, in some way, extra energy, such as heat, it will move to another orbit. An electron can only absorb the exact amount of energy required to raise its total energy to the energy value associated with another orbit; if given this amount of energy the electron will move to its new orbit and remain there until it loses some of its energy. The electron can only lose the exact amount of energy that will allow it to fall into a lower-energy orbit. This means that an electron can only absorb or lose energy in discrete amounts.

The simplified picture of the hydrogen atom given so far cannot account for all the observed phenomena, and it is necessary to extend the model by imagining that the electron can also move in elliptical orbits (Fig. 1.3). The number of elliptical orbits possible is equal to $n-1$, where n is the number of the basic circular orbits. The innermost orbit, $n=1$, has no elliptical paths associated with it; the next orbit, $n=2$, has a single elliptical path and so on. The circular orbits, numbers 1, 2, 3, etc. are said to form the K, L, M, etc. shells. The elliptical orbits are said to form sub-shells within these shells.

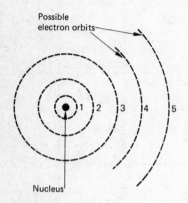

Fig. 1.2 Possible electron orbits in a hydrogen atom

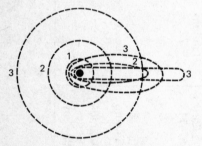

Fig. 1.3 Possible elliptical electron orbits in a hydrogen atom

Other Atoms

The extra-nuclear make-up of the other, more complex, elements can be deduced with accuracy up to the element of atomic number 18 (argon), by adding one more electron for each element in turn, bearing in mind that the number, x, of electrons permitted in a particular shell is given by the expression $x = 2n^2$, where n is the order of the shell. The innermost shell can only contain 2×1^2 or 2 electrons, the next shell 2×2^2 or 8 electrons, the next 2×3^2 or 18 electrons, and so on. The electrons in a particular shell follow paths of different eccentricities. Above atomic number 18 some gaps appear in this system because some orbits in the N shell have lower energy than some orbits in the M shell and are filled first.

The Periodic Table of the Elements is given in Table 1.1.

The atoms of all elements in group III have three electrons that are not part of a closed shell or sub-shell. Aluminium, for example, has 13 electrons, 10 of which completely fill the K and L shells; the M shell has only three electrons and is incomplete (since 18 electrons are necessary to fill it).

Indium is also in group III and has 49 electrons, 28 of which complete the K, L and M shells. Of the remaining 21 electrons, 18 completely fill three of the four sub-shells of the N shell and the remaining three enter the O shell (leaving the fourth sub-shell of the N shell unfilled). For both aluminium and indium, therefore, the extra-nuclear structure consists of a number of tightly bound closed shells and sub-shells with three electrons outside and not so tightly bound to the nucleus.

The electronic structure of all the other atoms is also in the form of a number of closed shells and sub-shells with a number of electrons in orbit outside. The nucleus of an atom plus the closed shells and sub-shells of electrons can be considered to be a positively charged central core, the positive charge being equal to $e \times n$, where e is the electronic charge and n is the number of electrons outside the central core. The number of electrons outside this central core is equal to the number of the group in the Periodic Table of the Elements to which the atom belongs. Thus all atoms in group I may be represented by the sketch of Fig. 1.4a, all atoms in group II by Fig. 1.4b, and so on.

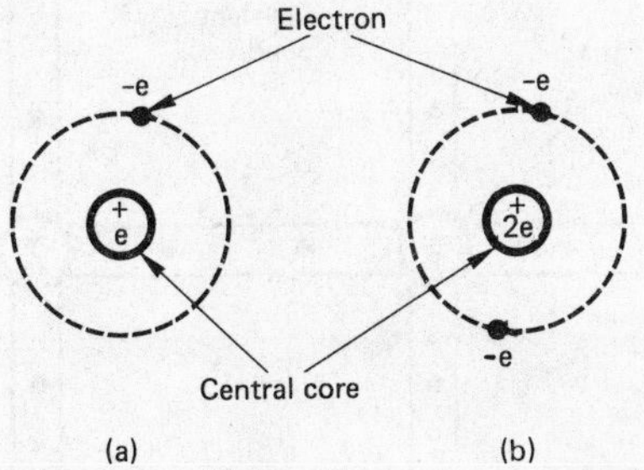

The outer electrons are known as **valence electrons** and determine the chemical properties of the element.

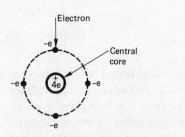

Fig. 1.4 Representation of (a) group I atoms and (b) group II atoms

Fig. 1.5 Representation of germanium or silicon atom

Intrinsic Semiconductors

The two semiconductor materials used in the manufacture of semiconductor devices, such as diodes and transistors, are germanium and silicon. It can be seen from Table 1.1 that both these materials fall into group IV of the Periodic Table of the Elements. An atom of either substance may be represented by a central core of positive charge $4e$ surrounded by four orbiting electrons each of which has a negative charge of e (Fig. 1.5). In the remainder of this chapter the discussion of semiconductors will be with reference to silicon but will apply equally well to germanium; any differences between the two materials will be mentioned in the appropriate places. In its solid state, silicon forms crystals of the diamond type, that is it forms a cubic lattice in which all the atoms (except those at the surface) are equidistant from their immediately neighbouring atoms. A study of crystal structures shows that the greatest number of atoms that can be neighbours to a particular atom at an equal distance away from that atom and yet be equidistant from one another is four. Hence each atom in a silicon crystal has four neighbouring atoms. In the crystal lattice each atom employs its four valence electrons to form covalent bonds with its four neighbouring atoms; each bond consisting of two electrons, one from each atom as shown in Fig. 1.6a. Each pair of electrons traverses an orbit around both its parent atom and a neighbouring atom. Each atom is effectively provided with an extra four electrons and these are sufficient to complete its final sub-shell. To simplify the drawings in the remainder of this chapter, covalent bonding will be represented in the manner of Fig. 1.6b.

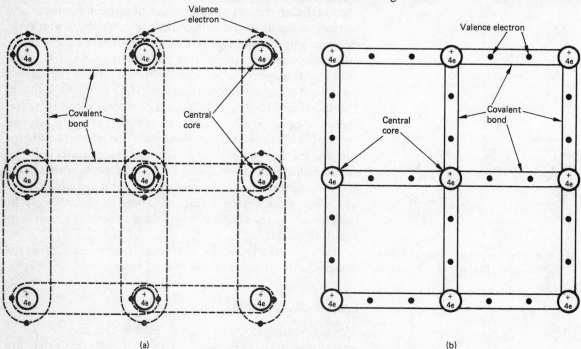

Fig. 1.6   Covalent bonding of atoms

If the temperature of the crystal is raised above absolute zero the lattice is thermally excited and some of the valence electrons receive sufficient energy to break free from a covalent bond. When this occurs the liberated electrons wander randomly in the crystal and are free to accept further energy from an applied electric field and contribute to electrical conduction. With further increase in temperature more covalent bonds are broken, and the conductivity of the silicon is increased because of the increased number of free electrons. This means that a semiconductor material has a negative temperature coefficient of resistance.

When an electron escapes from a covalent bond it leaves behind it an "absence of an electron" which, since it consists of a missing negative charge e, is equivalent to a positive charge of magnitude e. Such a positive charge is known as a HOLE. A hole exerts an attractive force on electrons and can be filled by a nearby passing electron that has been previously liberated from another broken covalent bond. This process is known as recombination and it causes a continual loss of holes and free electrons. At any given temperature the rate of recombination of holes and electrons is always equal to the rate of production of new holes and electrons so that the total number of free electrons and holes is constant.

When a covalent bond is broken it is said that a hole-electron pair has been created; both holes and electrons are known as charge carriers. The lifetime of a charge carrier is the time that elapses between its creation and its recombination with a charge carrier of opposite sign.

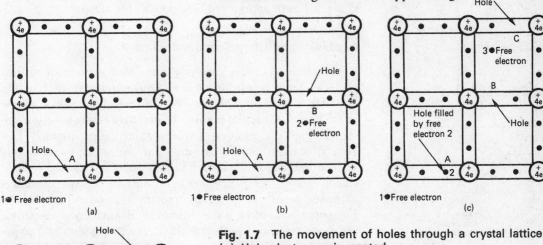

Fig. 1.7 The movement of holes through a crystal lattice
(*a*) Hole-electron pair created
(*b*) Second hole-electron pair created
(*c*) First hole disappeared and third hole-electron pair created
(*d*) Second hole disappeared

Movement of Holes through the Lattice

Fig. 1.7 shows a part of a silicon crystal in which the breaking of covalent bonds by thermal agitation of the lattice is taking place. In Fig. 1.7*a* thermal agitation of the crystal lattice has caused a covalent bond to break and produce a hole-electron pair at point A. An instant later a second hole-electron pair is produced at point B and two electrons are free to wander in the lattice (Fig. 1.7*b*). In Fig. 1.7*c* electron 2 has wandered close enough to the first hole to be attracted by its electric field and recombination has occurred. The original hole has apparently moved from position A to position B, but at the same time another hole-electron pair has been created at point C. Finally, in Fig. 1.7d, free electron 3 has travelled across the lattice and has recombined with the hole at point B and the effect is as though a hole has moved through the lattice from point A to point C.

The movement of both holes and electrons through the crystal is quite random but the holes appear to travel more slowly than do electrons. (This is because the movement of a hole in a particular direction actually consists of a series of discontinuous electron movements in the opposite direction.) If an electric field is set up in the crystal the electrons tend to *drift* in the direction of the field and the holes to *drift* in the opposite direction. Thus conduction of current in a pure semiconductor, known as intrinsic conduction, takes place (current flow is conventionally in the opposite direction to electron flow). Intrinsic conduction increases with increase in temperature at the approximate rate of 5% per degree Centigrade for germanium and 7% per °C for silicon.

Extrinsic (Impurity) Semiconductors

If an extremely small, carefully controlled amount of an impurity element is introduced into a silicon crystal, each of the impurity atoms will take the place of one of the silicon atoms in the lattice. Since the number of impurity atoms is very much smaller than the number of silicon atoms (approximately 1 in 10^8), it is reasonable to assume that the lattice is essentially undisturbed and that each impurity atom is surrounded by four silicon atoms. In practice, the impurities are always substances in either group III or group V of the Periodic Table of the Elements, and have either three or five valence electrons. Elements typically employed are arsenic, antimony and phosphorus in group V, and indium, aluminium and gallium in group III. The process of introducing impurity atoms into a silicon crystal is called "doping" and a treated crystal is said to be "doped".

n-Type Semiconductor

Suppose a silicon crystal has been doped with a small quantity of phosphorus, a substance having five valence electrons. Each phosphorus atom will set up covalent bonds with its four neighbouring atoms but, since only four of its valence electrons are required for this purpose, a spare electron exists (Fig. 1.8). This surplus electron is not bound to its parent atom and is free to wander in the lattice.

A free electron is created in the silicon crystal lattice for each impurity atom introduced without the creation of corres-

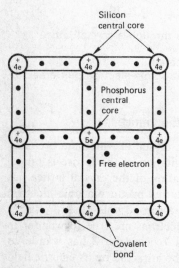

Silicon central core

Phosphorus central core

Free electron

Covalent bond

Fig. 1.8 Lattice of n-type silicon crystal

ponding holes. Hole-electron pairs are, however, still produced by thermal agitation of the lattice. The number of free electrons in the crystal is much greater than the number of holes, negative charges predominate and so the crystal is said to be n-type. Since each impurity atom donates a free electron to the crystal, the impurity atoms are known as DONOR atoms.

p-Type Semiconductor

If, instead of phosphorus, a group III element such as boron is introduced into a silicon crystal, each boron atom will attempt to form a covalent bond with each of its four neighbouring silicon atoms. Boron, however, has only three valence electrons and so only three of the bonds can be completed (Fig. 1.9). One hole is introduced into the lattice for each impurity atom and is able to move about in the crystal in the same way as a hole produced by thermal agitation. In this case holes are in the majority and the material is known as p-type, while the impurity atoms are called ACCEPTOR atoms.

A crystal of n-type or p-type silicon is electrically neutral because each impurity atom introduced into the lattice is itself neutral. In n-type material, electrons are the MAJORITY CHARGE CARRIERS and holes are the MINORITY CHARGE CARRIERS. In p-type material the electrons are the minority charge carriers and holes are the majority charge carriers.

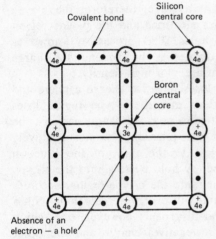

Fig. 1.9 Lattice of p-type silicon crystal

Current Flow

If a potential difference is maintained across an extrinsic semiconductor (Fig. 1.10), a drift current will flow into the material at one end and out of the material at the other. The positively charged central cores cannot move from their positions in the crystal lattice and so the current flowing *into* the material can only consist of electrons flowing *out* and the current flowing *out* of the material is actually an inward flow of electrons.

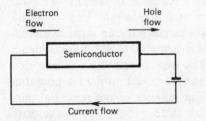

Fig. 1.10 Current flow in an extrinsic semiconductor

The p-n Junction

If a silicon crystal is doped with donor atoms at one end and acceptor atoms at the other the crystal will have both p-type and n-type regions and there will be a junction between them. In Fig. 1.11 the plane AA′ is the p-n junction; only the free electrons and holes have been shown, to clarify the drawing. Both regions include charge carriers of either sign but in the n-type region electrons are in the majority and in the p-type

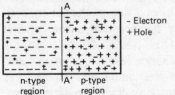

Fig. 1.11 The formation of a p-n junction

region holes predominate. In both regions the probability of a minority charge carrier meeting and recombining with a majority charge carrier is high and the minority charge carrier lifetime is short.

The free electrons and holes have completely random motions and wander freely in the lattice. However, since there are more electrons to the left of the p-n junction than to the right and more holes to the right of the junction than to the left, *on average* more electrons cross the junction from left to right than from right to left, and more holes cross from right to left than from left to right. On average, therefore, the n-type region gains holes and loses electrons and the p-type region gains electrons and loses holes. This process is known as DIFFUSION and may be defined as the tendency for charge carriers to move away from areas of high density.

Since the n-type region loses negative charge carriers and gains positive charge carriers, and the p-type region loses positive charge carriers and gains negative charge carriers, the region just to the left of the junction becomes positively charged and the region just to the right of the junction becomes negatively charged. A hole passing into the n-type region, or an electron passing into the p-type region, becomes a minority charge carrier and will probably recombine with a carrier of opposite sign and disappear; however, one region has still lost a positive (or negative) charge and the other region has gained a positive (or negative) charge. The movement of holes and electrons across the junction constitutes a current and this is known as the DIFFUSION CURRENT.

If the crystal was neutral before diffusion took place it must be neutral afterwards. Further, because both regions were also originally neutral they must contain equal and opposite charges after diffusion. These charges have an attractive electric force between them and are not able to diffuse away from the vicinity of the junction. The two charges are concentrated immediately adjacent to the junction, and produce a potential barrier across the junction. The polarity of the potential barrier is such as to oppose the further diffusion of majority charge carriers across the junction, but to aid the movement of minority charge carriers; this gives rise to a minority charge carrier current in the opposite direction to the diffusion current.

The difference in potential from one side of the junction to the other is called the *height of the potential barrier* and is measured in volts. The height of the potential barrier attains such a value that the majority charge carrier (diffusion) and minority charge carrier currents are equal and so the net current across the junction is zero. Any charge carriers entering the region on either side of the junction over which the barrier potential is effective are rapidly swept out of it, and hence this region is depleted of charge carriers. The DEPLE-

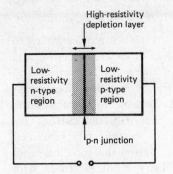

Fig. 1.12 The unbiased p-n junction

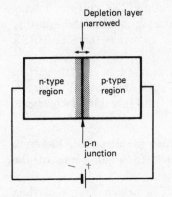

Fig. 1.13 The forward-biased p-n junction

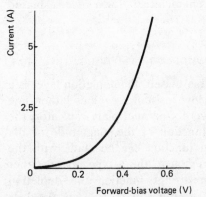

Fig. 1.14 The current/voltage characteristic of a forward biased p-n junction

TION LAYER, as it is called, is a region of relatively high resistivity and is approximately 0.001 mm in width.

If an external source of e.m.f. is applied across the p-n junction, the equilibrium state of the junction is disturbed and the potential barrier is either increased or decreased according to the polarity of the external e.m.f. The silicon crystal consists of two regions of low resistivity separated by a region of high resistivity, the depletion layer, and the application of an e.m.f. across the crystal is effectively the same as placing it across the depletion layer (see Fig. 1.12).

The Forward-biased p-n Junction

If a battery is connected across the crystal in the direction shown in Fig. 1.13, holes are repelled from the positive end of the crystal and are caused to drift towards the junction; and electrons are repelled from the negative end of the crystal and also drift towards the junction. This drift of holes and electrons towards the junction reduces both the width of the depletion layer and the height of the potential barrier, and the junction is said to be FORWARD BIASED. The reduction in the height of the potential barrier allows majority charge carriers of lower energy to cross the junction and, since the minority charge carrier current remains constant, there is a net majority charge carrier current across the junction from the p-type region to the n-type region. This current increases very rapidly with increase in the forward bias voltage as can be seen from the typical current/voltage characteristic shown in Fig. 1.14.

The holes drifting through the p-type region towards the p-n junction may be considered to have been injected by the positive terminal of the battery. Some of these holes may recombine with electrons diffusing across the junction in the other direction and so the hole current across the junction is slightly less than the injected hole current. After they have passed across the junction the holes recombine with the excess electrons in the n-type region. Similarly, the negative battery terminal injects electrons into the n-type region and most of these electrons cross the junction. The total current is the sum of the electron and hole currents and is constant throughout the crystal. The current enters the p-type region as a hole current and leaves the n-type region as an electron current, i.e. in the forward direction current flow is by majority charge carriers.

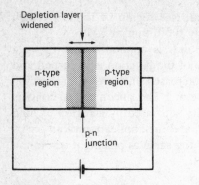

Fig. 1.15 The reverse-biased p-n junction

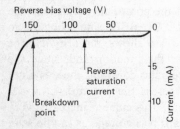

Fig. 1.16 The current/voltage characteristic of a reverse-biased p-n junction

The Reverse-biased p-n Junction

Fig. 1.15 shows a p-n junction biased in such a direction as to attract majority charge carriers away from the junction and thus to increase both the height of the potential barrier and the width of the depletion layer. Fewer majority charge carriers now have sufficient energy to be able to surmount the potential barrier and the majority charge carrier current decreases. The minority charge carrier current has remained constant and so a net current flows across the junction from n-type region to p-type region. This current increases with increase in the reverse bias voltage until the point is reached where almost no majority charge carriers possess sufficient energy to be able to cross the junction. The current flowing across the junction is then constant and equal to the minority charge carrier current and it is then known as the REVERSE SATURATION CURRENT.

If the reverse bias voltage is increased beyond a certain value a rapid increase in current occurs; this critical voltage is the BREAKDOWN VOLTAGE of the junction. Two effects are responsible for breakdown:

(a) the ZENER EFFECT in which the electric field across the junction is strong enough to break some of the covalent bonds and

(b) the AVALANCHE EFFECT in which charge carriers are accelerated to such an extent that they are able to break covalent bonds by collision.

A typical current/voltage characteristic for a reverse-biased p-n junction is shown in Fig. 1.16.

The Capacitance of a p-n Junction

When a p-n junction is reverse-biased, the depletion layer is a high-resistance region with low-resistance regions either side and so it acts as though it were a parallel-plate capacitor, the capacitance of which is a function of the magnitude of the applied bias voltage. A p-n junction can be made with the transition from the p-type region to the n-type region either abrupt or gradual. For an abrupt junction the depletion capacitance is proportional to the square root of the bias voltage, and for a gradual junction, to the cube root.

Exercises
1.1. By reference to the formation of a potential barrier and a current/voltage characteristic, explain the rectifying action of a p-n junction.
1.2. Describe, with the aid of suitable sketches, how a hole appears to move in a semiconductor material.

1.3. How does an increase in the temperature of a semiconductor material affect its intrinsic conduction? The resistivity of silicon is 2300 ohm-metres at 27 °C. What will be its resistivity if the temperature is increased by (i) 10% and (ii) 25%? Assume that the intrinsic conductivity of silicon increases by 7% per °C increase in temperature.

1.4. Explain clearly why the current flowing across a p-n junction is the sum of the hole and the electron currents flowing across the junction. In which direction does the electron current flow?

1.5. Answer the following questions relating to current flow across an unbiased p-n junction.
(a) Why do electrons diffuse from the n-type region to the p-type region?
(b) Does the potential barrier assist or oppose such diffusion?
(c) Why does the flow of electrons across the junction not produce a net flow of current?
(d) What happens to the holes and electrons that are produced in both regions by thermal agitation of the crystal lattice?

1.6. (a) Draw a sketch of an unbiased p-n junction and mark on your sketch (i) the position of the depletion layer, (ii) the majority charge carriers in each region, and (iii) the current flowing.
(b) Redraw the junction with an applied forward bias voltage and show the effect on the depletion layer. Indicate, with arrows, the directions of the majority charge carrier and minority charge carrier currents, and the total current in the circuit. How would the total current be measured?

1.7. (a) Explain why the resistance of a conductor increases, but the resistance of a semiconductor decreases, with increase in temperature. Do any other differences between a conductor and a semiconductor exist?
(b) Name two conductor materials and two semiconductor materials in common use in telecommunications work.

1.8. What is meant by the term *intrinsic semiconductor*? Briefly explain how an intrinsic semiconductor material can be made into either an n-type or a p-type extrinsic semiconductor. What is meant by majority charge carriers and minority charge carriers in a p-type region? Is the number of either or both of these carriers dependent upon temperature?

Short Exercises

1.9. Define the properties of a semiconductor material in relation to insulators and conductors.

1.10. Explain the difference between diffusion current and drift current in a semiconductor material.

1.11. Draw a reverse-biased p-n junction and show clearly the direction of the current flowing in the external circuit.

1.12. Draw a forward-biased p-n junction and indicate the current flowing.

1.13. What is meant by the potential barrier of a p-n junction? Is it raised or lowered by a forward-bias voltage? Why is the current across an unbiased p-n junction equal to zero?

1.14. What is meant by the terms majority charge current and minority charge current? Are holes or electrons minority charge carriers in (a) the p-type region and (b) the n-type region?

1.15. Explain why a p-n junction possesses capacitance. Is this capacitance increased or decreased when the reverse bias is reduced?

1.16. Is a semiconductor negatively charged when it is doped with donor atoms? Give reasons for your answer.

1.17. Describe two non-electrical examples of diffusion.

1.18. What is meant by the following terms used in conjunction with semiconductors: donor atom, recombination, lifetime, hole-electron pair?

1.19. An atom has 11 electrons orbiting a nucleus containing twelve neutrons and 11 protons. What is the mass number of the atom? If the element concerned is in group 1 of the periodic table of the elements how many electrons are there in the outermost orbit? How many electrons are there in the inner-most orbit? What name is given to the electrons in the outer orbit?

1.20. In a p-type semiconductor material holes are present in much greater number than are electrons. Why, therefore, is the semiconductor electrically neutral?

2 Semiconductor Diodes

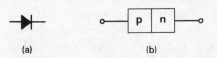

Fig. 2.1 The basic semiconductor diode

Construction

The semiconductor diode is a device that has a high resistance to the flow of current in one direction and a low resistance in the other. The diode is widely employed in electronic circuitry for many different purposes and it consists essentially of a p-n junction formed in either a silicon or a germanium crystal (Fig. 2.1*b*). The symbol for a semiconductor diode is shown in Fig. 2.1*a*.

The direction in which the diode offers little opposition to current flow is indicated by the arrowhead. The semiconductor diode possesses a number of advantages over the thermionic valve diode; it does not require a heater supply, it is much smaller and lighter, and it is very much more reliable.

Germanium and silicon for use in the manufacture of semi-conductor diodes must be first purified until an impurity concentration of less than 1 part in 10^{10} is achieved. The wanted inpurity atoms, donors and/or acceptors, are then added in the required amounts and the material is made into a single crystal.

A p-n junction may be formed in a number of different ways but two basic techniques are generally employed, either singly or in combination. An example of the first method is outlined in Fig. 2.2 and consists of alloying an indium pellet on to an n-type germanium wafer.

To make the n-type germanium wafer some intrinsic germanium and a small amount of impurity are melted in a crucible in a vacuum, and a seed crystal is lowered into the melt to a depth of a few millimetres. The temperature of the molten germanium is just above the melting point of the seed crystal, and the few millimetres of seed immersed in the melt also melt. The seed is rotated at a constant velocity and at the same time is slowly withdrawn from the melt, thus forming an

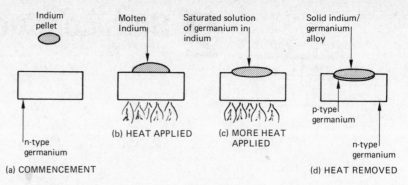

Fig. 2.2 The "alloying" method of forming a p-n junction

n-type crystal. By careful control of the process the required impurity concentration can be achieved.

A pellet of indium is placed on the germanium wafer and is heated to a temperature above the melting point of indium but below the melting point of germanium. The indium melts and dissolves the germanium until a saturated solution of germanium in indium is obtained. The wafer is then slowly cooled and in the cooling a region of p-type germanium is produced in the wafer, and an alloy of germanium and indium (mainly indium) is deposited on the wafer. A silicon alloyed p-n junction can be formed using the same method but with aluminium as the acceptor element.

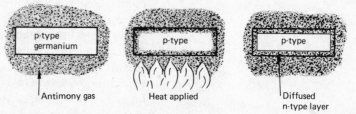

Fig. 2.3 The "diffusion" method of forming a p-n junction

The second method of producing a p-n junction to be considered here is diffusion and it is outlined in Fig. 2.3. The p-type germanium is heated to a temperature very nearly equal to the melting point of germanium, and is surrounded by the donor element antimony in gaseous form. The antimony atoms will diffuse into the germanium to produce an n-type region. If an n-type germanium crystal is used gallium is employed, in gaseous form, as the acceptor element to produce a p-type region in the crystal. When a silicon device is to be manufactured, boron is used as the acceptor element and phosphorus as the donor element.

A JUNCTION DIODE consists of a crystal having both p-type and n-type regions. Junction diodes are made from either germanium or silicon, the former having the advantage of a lower forward resistance and the latter the advantages of a

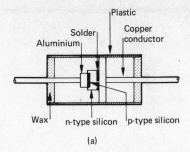

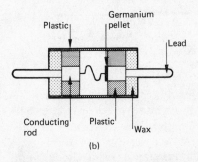

Fig. 2.4 Construction of (a) a silicon junction diode and (b) a germanium point-contact diode

higher breakdown voltage and a lower reverse saturation current. Connection to the junction is made by wires fixed to each of the two regions. The complete device is usually enclosed in a hermetically sealed container to prevent the entry of moisture (see Fig. 2.4a).

Germanium POINT-CONTACT DIODES are also employed; a typical construction is shown in Fig. 2.4b. A point-contact diode consists essentially of a pellet of n-type germanium that has the tip of a tungsten wire, or whisker, pressing on to its surface. Connection to the whisker is via two copper leads. During the manufacture of a point-contact diode, a pulse of current is passed through the diode and this results in the area of the pellet immediately adjacent to the tip of the whisker becoming a p-type region. A low capacitance p-n junction is then produced in the pellet.

Semiconductor Diode Current/Voltage Characteristics

The current/voltage characteristic of a semiconductor diode is a graph of the current flowing in the device plotted against the voltage applied across it.

It can be measured with the aid of the circuit arrangement of Fig. 2.5. With the switch in the position shown, the diode is reverse-biased; to forward bias the diode, the switch is thrown to

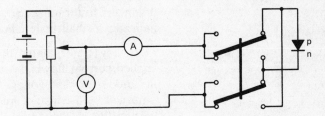

Fig. 2.5 Circuit for measuring the current/voltage characteristic of a semiconductor diode

its other position in order to reverse the polarity of the applied voltage. For each position of the switch, the applied voltage is increased from zero in a number of steps and the current flowing at each step is noted. The noted current values are then plotted to a base of voltage.

Typical current/voltage characteristics for silicon and germanium diodes are shown in Fig. 2.6.

Note that the forward current does not increase to any noticeable extent until the forward bias voltage is greater than about 0.6 V for the silicon diode and about 0.2 V for the germanium diode. The other features of importance are (i) the reverse saturation current, and (ii) the reverse breakdown

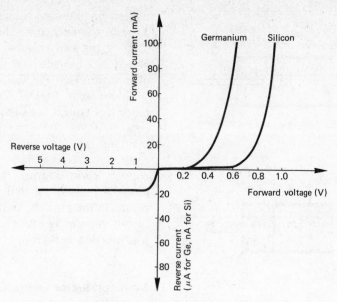

Fig. 2.6 Current/voltage characteristics for (a) a silicon diode and (b) a germanium diode

voltage (not shown). The a.c. resistance of a diode at a particular d.c. voltage is equal to the reciprocal of the slope of the characteristic at that point, that is

$$r_{ac} = \frac{\text{Change in voltage}}{\text{Resulting change in current}} = \frac{\delta V}{\delta I} \text{ ohm}$$

(*Note* The Greek letter δ (delta) means "a change of" wherever it appears in formulae. So, δt is a change of time. Generally it indicates a small-scale change.)

At any point along the characteristic the ratio voltage applied/current flowing is a measure of the d.c. resistance of the diode for that voltage. If the characteristic is linear this ratio will be a constant quantity but should the characteristic be non-linear, the d.c. resistance will vary with the point of measurement.

EXAMPLE 2.1

Calculate the a.c. resistance of the semiconductor diode whose characteristic is shown in Fig. 2.7 at the point +1V.

Solution
The a.c. resistance r_{ac} is equal to $\delta V/\delta I$ and to find the forward resistance at the point $V = 1$ V it is necessary to select two equidistant points either side of this voltage and then, by projection to and from the characteristic, find the corresponding values of current.

Two points 0.2 V either side of +1 V have been selected, hence $\delta V = 0.4$ V.

Projection upwards from these points to the curve and then from the curve to the current axis, as shown by the dotted lines, shows that the corresponding values of current are 15.5 mA and 5.5 mA, i.e.

$$\delta I = 10 \text{ mA} \qquad r_{ac} = 0.4/10 \times 10^{-3} = 40 \text{ ohms} \qquad (Ans.)$$

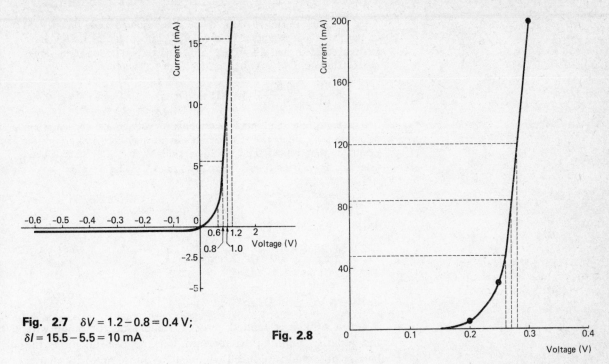

Fig. 2.7 $\delta V = 1.2 - 0.8 = 0.4\,\text{V}$; $\delta I = 15.5 - 5.5 = 10\,\text{mA}$

Fig. 2.8

The slope of the reverse saturation current curve is very small and cannot be measured from the characteristic; this means that the reverse a.c. resistance of the diode is high, of the order of several thousands of ohms.

EXAMPLE 2.2

The current/voltage characteristic of a semiconductor diode is given in the table.

Voltage (V)	0.05	0.10	0.15	0.20	0.25	0.30
Current (mA)	0.2	0.4	0.6	4.0	30	200

Plot the characteristic and use it to determine (i) the d.c. resistance and (ii) the a.c. resistance of the diode at the point when $V = 0.27$ volts.

Solution
The current/voltage characteristic of the diode is shown plotted in Fig. 2.8.

(i) The d.c. resistance of the diode at the point $V = 0.27\,\text{V}$ is found by drawing a line upwards from the voltage axis to the characteristic and then projecting on to the current axis. The corresponding d.c. current value is 84 mA and therefore

$$r_{dc} = \frac{V}{I} = \frac{0.27}{84 \times 10^{-3}} = 3.21\,\Omega \quad (Ans.)$$

(ii) The a.c. resistance of the diode is determined using the method of the preceding example. Points 0.01 V either side of 0.27 V have been selected so that $\delta V = 0.02$ V. The corresponding current values are 120 mA and 48 mA; hence $\delta I = 72$ mA. Therefore

$$r_{ac} = \frac{\delta V}{\delta I} = \frac{0.02}{72 \times 10^{-3}} = 0.28 \, \Omega \qquad (Ans.)$$

It is important that small increments of voltage are chosen when calculating r_{ac}, otherwise considerable error may occur. Suppose, for example, that points 0.03 V either side of $V = 0.27$ V had been selected. Then $\delta V = 0.06$ V and $\delta I = 178$ mA giving

$$r_{ac} = \frac{0.06}{178 \times 10^{-3}} \quad \text{or} \quad 0.34 \, \Omega$$

This is a percentage error of

$$\frac{0.34 - 0.28}{0.28} \times 100 \quad \text{or} \quad 21.43\%$$

Forward Voltage Drop

When a voltage is applied across a semiconductor diode with the polarity required to forward bias its p-n junction, the barrier potential is reduced which allows more majority charge carriers to cross the junction. Since the minority charge carrier current is unaffected, a net current flows across the junction. As the forward bias voltage is increased, the current is very small at first but it increases rapidly once the voltage has exceeded a particular threshold value. For a silicon p-n junction this threshold voltage is approximately 0.6 V, but it is only about 0.2 V for a germanium junction. This is clearly shown by the typical characteristics of Fig. 2.6.

Reverse Saturation Current

When a reverse bias voltage is applied to a semiconductor diode, the barrier potential is increased and fewer majority charge carriers have sufficient energy to cross the junction. With increase in the reverse bias voltage, the point is reached where the current consists almost entirely of minority charge carriers. The current flowing then becomes more or less constant and is known as the REVERSE SATURATION CURRENT. The reverse saturation current in a germanium diode is very much greater than the reverse saturation current of a silicon diode of comparable maximum forward current rating. Thus, for the smaller types of diode the reverse saturation current would be a few microamperes in a germanium diode but only a few nanoamperes in a silicon diode.

If the temperature of the p-n junction is increased, further hole electron pairs will be produced and the reverse saturation current will become larger.

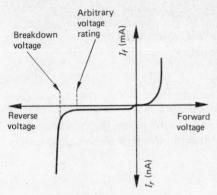

Fig. 2.9 Voltage breakdown in a semiconductor diode

Breakdown Voltage

If the reverse bias voltage applied to a diode is steadily increased, the current will remain at an approximately constant value until a point is reached where a sudden and large increase in current takes place (Fig. 2.9). In the breakdown region the reverse resistance of the diode will be low. Also this large increase in current will dissipate power within the diode and may lead to the destruction of the device. It is necessary therefore, to ensure that the diode is not driven into its breakdown region. An arbitrary voltage rating is determined and quoted by the manufacturer for each type of diode which, if not exceeded, will ensure the satisfactory working of the device. The PEAK INVERSE VOLTAGE (p.i.v.) varies considerably with the type of diode and may easily be a few hundreds of volts.

EXAMPLE 2.3

A silicon junction diode has the following parameters: maximum forward current 250 mA, forward voltage drop of 1.2 V at a forward current of 30 mA, a maximum reverse voltage of 50 V, and a reverse saturation current of 0.05 μA at the maximum reverse voltage.
   (i) Sketch the current/voltage characteristic of the diode.
   (ii) Determine its d.c. resistance at a forward voltage of 1.2 V.

Solution
(i) The required current/voltage characteristic is given in Fig. 2.10.
(ii) The d.c. resistance at $V = 1.2$ V is

$$r_{dc} = \frac{1.2}{0.3} = 4 \, \Omega$$

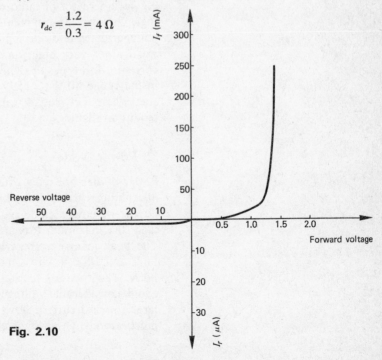

Fig. 2.10

Types of Diode and their Application

The important parameters of semiconductor diodes are

(1) Forward and reverse a.c. resistances
(2) Maximum forward current
(3) Junction capacitance
(4) Behaviour in breakdown region.

Depending upon the intended application of a diode, one or more of these parameters may be of prime importance.

The main types of diode used in modern electronic circuitry are

(1) Signal diodes
(2) Power diodes
(3) Zener diodes
(4) Varactor diodes

(1) Signal Diodes

The term *signal diode* include all diodes which have been designed for use in circuits where large current and/or voltage ratings are not required. The usual requirements are for a large reverse resistance/forward resistance ratio and minimum junction capacitance. Some of the commerically available signal diodes are listed as general-purpose types while others are best suited to a particular circuit application, e.g. as a detector of radio waves, or as an electronic switch in logic circuitry. The maximum reverse voltage, or peak inverse voltage, that the diode is likely to be called upon to handle is usually not very high, and neither is the maximum forward current. Most types of signal diode have a peak inverse voltage in the range 30 V to 150 V and a maximum forward current somewhere between 40 and 250 mA, but higher values are readily available.

(2) Power Diodes

Power diodes are most often employed for the conversion of alternating current into direct current, i.e. as rectifiers. The important power diode parameters are the peak inverse voltage, the maximum forward current, and the resistance ratio. The peak inverse voltage is likely to be somewhere between 50 V and 1000 V with a maximum forward current of perhaps 30 A. The forward resistance must be as low as possible to avoid considerable voltage drop across the diode when the large forward current flows; this resistance is usually not very much more than an ohm or two.

(3) Zener Diodes

The large reverse current which flows when the breakdown voltage of a diode is exceeded need not necessarily result in damage to the device.

A zener diode is fabricated in a way which allows it to be operated in the breakdown region without damage, provided the current is restricted by external resistance to a safe value. The large current at breakdown is brought about by two factors, known as the zener and the avalanche effects. At voltages up to about 5 V the electric field near to the junction is strong enough to pull electrons out of the covalent bonds holding the atoms together. Extra hole-electron pairs are produced and these are available to augment the reverse current. This is known as the *zener effect*.

The *avalanche effect* occurs if the reverse bias voltage is made larger than 5 V or so. The velocity with which the charge carriers move through the crystal lattice is increased to such an extent that they attain sufficient kinetic energy to *ionize* atoms by collision. An atom is said to have been ionized when one of its electrons has been removed. The extra charge carriers thus produced travel through the crystal lattice and may also collide with other atoms to produce even more carriers by ionization. In this way the number of charge carriers, and hence the reverse current, is rapidly increased.

Zener diodes are available in a number of standardized *reference voltages*. For example, it is possible to obtain a zener diode with a reference (breakdown) voltage of 8.2 V. An alternative name for the device is the *voltage reference diode*. The most common application of the zener diode is in the voltage stabilizing circuits which are discussed in Chapter 9. It is also employed as a voltage reference.

EXAMPLE 2.4

A zener diode is advertised as having a breakdown voltage of 20 V with a maximum power dissipation of 400 mW.

What is the maximum current the diode should be allowed to handle?

$$I = P/V = 0.4/20 = 20 \, \text{mA} \qquad (Ans.)$$

(4) Varactor Diodes

A p-n junction is a region of high resistivity sandwiched in between two regions of relatively low resistivity. Such a junction therefore possesses capacitance, the magnitude of which is given by

$$C = \frac{\varepsilon A}{W} \qquad (2.1)$$

where ε is the permittivity of the semiconductor material, A is the area of the junction, and W is the width of the depletion layer. W is not a constant quantity but, instead, varies with the magnitude and the polarity of the voltage applied across the junction.

Most semiconductor diodes are manufactured in such a way that their junction capacitance is minimized, but a varactor diode has been designed to have a particular range of capacitance values.

The varactor diode is operated with a reverse bias and then its junction capacitance is inversely proportional to the square root of the bias voltage V, i.e.

$$C = \frac{K}{\sqrt{V}} \qquad (2.2)$$

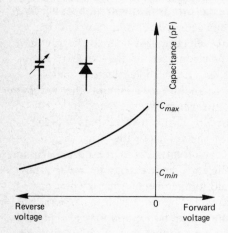

Fig. 2.11 shows graphically how the capacitance of a varactor diode varies with the reverse bias voltage, and it also shows the symbol for a varactor diode. Typically, the capacitance variation might be 2–12 pF, or 20–28 pF, or perhaps 27–72 pF.

Fig. 2.11 Varactor diode characteristics

EXAMPLE 2.5

A varactor diode has a capacitance of 5 pF when the reverse bias voltage applied across it is 4 V. Determine the diode capacitance if the bias voltage is increased to 6 V.

Solution

From equation (2.2) $\quad 5 = \dfrac{K}{\sqrt{4}} \quad$ i.e. $K = 10$ pF

Therefore, when the voltage has increased to 6 V,

$\quad C = 10/\sqrt{6} = 4.082$ pF $\qquad$ (*Ans.*)

Exercises

2.1. A semiconductor diode has the following data

Forward voltage (V)	0	0.1	0.2	0.3	0.4	0.5	0.6
Forward current (A)	0	0.01	0.5	4	12	30	74

The reverse saturation current is 15 μA and the reverse breakdown voltage is 60 V. Plot the static characteristic of the diode. Is this a silicon or a germanium diode? Estimate a suitable value for the maximum reverse voltage.

2.2. A semiconductor diode has the data given.

Forward voltage (V)	0	0.2	0.4	0.5	0.6	0.7	0.8	0.9
Forward current (mA)	0	0	0.02	0.2	1	8	20	60

The reverse saturation current is 20 nA and the maximum reverse voltage is 50 V. Plot the static characteristic of the diode. Is this a silicon or a germanium diode? Estimate the breakdown voltage of the diode. Suggest a use for the diode.

2.3. A semiconductor diode has the data given.

Forward voltage (V)	0	0.4	0.8	1.0	1.2	1.4	1.6	1.8	2.0
Forward current (A)	0	0.03	0.06	0.25	1	6	14	28	70

At the maximum reverse voltage of 300 V the reverse saturation current is 1 μA. Plot the static characteristic of the diode and determine (i) its d.c. forward resistance when the forward voltage is 1 V, and (ii) the a.c. resistance at the point $V = 1.4$ V. What kind of diode is this?

2.4. A zener diode has the reverse voltage characteristic given by the data in the table.

Reverse voltage (V)	-1	-3	-5	-5.6	-5.7	-5.8	-5.9	-6.0
Reverse current (mA)	0.01	0.01	0.02	1	13	25	37.5	50

Plot the reverse characteristic of the diode. What is the breakdown voltage of the diode? Determine the a.c. resistance of the diode in its breakdown region.

2.5. The quoted parameters of a germanium signal diode are as follows: maximum reverse voltage 110 V; maximum forward current 150 mA; forward voltage drop of 2 V when a forward current of 30 mA is flowing; reverse saturation current of 70 μA when the reverse voltage is 100 V.
  (i) Sketch the static characteristic of the diode.
  (ii) Estimate the breakdown voltage of the diode.
  (iii) Calculate the a.c. resistance of the diode for a forward current of 50 mA.

2.6. What is a varactor diode? A varactor diode has the capacitance/voltage characteristic given by the data.

Reverse voltage (V)	-1	-2	-3	-4	-5	-6
Diode capacitance (pF)	7.5	5.3	4.33	3.7	3.35	3.1

Plot a graph of diode capacitance against reverse bias voltage and hence determine the diode capacitance when the bias voltage is 3.5 V.

2.7. Briefly explain the meaning of the term *reverse saturation current* used in conjunction with semiconductor diodes. The reverse saturation current of a germanium diode is 10 μA at a temperature of 25°C. Calculate the expected value of the current if the temperature should increase to 41°C.

2.8. Briefly explain the mechanisms causing breakdown in a zener diode. A zener diode has a breakdown voltage of 7.5 V and a maximum power dissipation of 1 W. Determine the maximum current the diode can pass without damage.

Short Exercises

2.9. Explain why the peak inverse voltage of a semiconductor diode is an important parameter.

2.10. Diodes are to be selected for the applications listed. Complete the table.

Application	Diode Type	Application	Diode Type
Voltage-tunable capacitance		Rectifier unit	
Gates in logic circuitry		Voltage stabilization	
Reference voltage		Detector of radio waves	

2.11. Sketch, on the same axes, typical static characteristics for germanium and silicon diodes. Label clearly the values of forward voltage drop and reverse saturation current.

2.12. What is the effect on the reverse saturation current of (i) a silicon diode and (ii) a germanium diode of an increase in temperature? Briefly outline the reasons for your answer.

2.13. Explain why a p-n junction possesses capacitance. Why can this capacitance be varied by means of the bias voltage applied to the junction? Why is a varactor diode operated with a reverse bias?

2.14. List the donor and acceptor atoms used to form p-type and n-type regions in germanium and in silicon, using (i) the alloying and (ii) the diffusion technique of manufacture.

2.15. Explain how you would measure the current/voltage characteristic of a semiconductor diode. Do not use the circuit given in Fig. 2.5.

3 Transistors

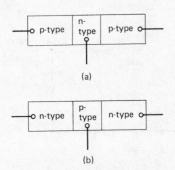

Fig. 3.1 (a) a p-n-p transistor, (b) an n-p-n transistor

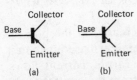

Fig. 3.2 Symbols for (a) a p-n-p transistor and (b) an n-p-n transistor

Types of Transistor

The transistor is a semiconductor device that can amplify an electrical signal, act as an electronic switch, and perform a number of other functions. Basically a transistor consists of a germanium or silicon crystal containing three separate regions. The three regions may consist of either two p-type regions separated by an n-type region (Fig. 3.1a) or two n-type regions separated by a p-type region (Fig. 3.1b). The first type of transistor is known as a p-n-p transistor and the second type as an n-p-n transistor. Both types of transistor are employed, sometimes together in the same circuit, but the discussion in this chapter will be in terms of the p-n-p transistor. However, for the corresponding operation of an n-p-n transistor it is merely necessary to read electron for hole, hole for electron, negative for positive, and positive for negative.

The middle of the three regions in a transistor is known as the BASE and the two outer regions are known as the EMITTER and the COLLECTOR. In most transistors the collector region is made physically larger than the emitter region because it will be expected to dissipate a greater power. The symbol for a p-n-p transistor is given in Fig. 3.2a and the symbol for an n-p-n transistor in Fig. 3.2b. Note that the emitter lead arrowhead is pointing in different directions in the two figures, pointing inwards for the p-n-p transistor and outwards for the n-p-n transistor. It will shortly become evident that the arrowhead indicates the direction in which holes travel in the emitter.

Both p-n-p and n-p-n transistors are generally classified into one of the following groups:

(a) small-signal low-frequency
(b) low-power and medium-power low-frequency
(c) high-power low-frequency

(d) small-signal high-frequency
(e) switching

The majority of the transistors listed in manufacturer's/distributor's catalogues are silicon types.

The Action of a Transistor

A p-n-p transistor contains two p-n junctions and is normally operated so that one junction, the EMITTER/BASE JUNCTION, is forward-biased and the other, the COLLECTOR/BASE JUNCTION, is reverse-biased. This is shown in Fig. 3.3 together with the directions of the various currents that flow in the transistor. The usual convention whereby the direction of current flow is opposite to the direction of electron movement has been employed.

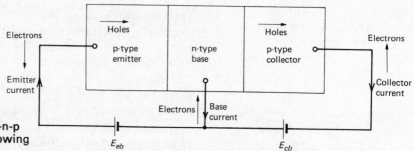

Fig. 3.3 Bias voltages for a p-n-p transistor and the currents flowing

Consider that, initially, the emitter/base bias voltage E_{eb} is zero. Then the majority charge carrier current crossing the emitter/base junction is equal to the minority charge carrier current that is flowing in the opposite direction and the net junction current is zero. The collector/base junction is reverse-biased by the bias voltage E_{cb} and so a small minority charge carrier current flows in the collector lead. This current is the reverse saturation current discussed in the previous chapter but now it is known as the COLLECTOR LEAKAGE CURRENT and is given the symbol I_{CBO}.

If the emitter/base bias voltage is increased in the positive direction by a few tenths of a volt, the emitter/base junction is forward-biased and a majority charge carrier current flows. This current consists of holes travelling from the emitter to the base and electrons passing from the base to the emitter. Only the hole current is useful to the action of the transistor, as will soon be evident, and it is therefore made much larger than the electron current by doping the base much more lightly than the emitter. The ratio of the hole current to the total emitter current is known as the *emitter injection ratio* or the emitter efficiency, symbol γ. Typically, γ is approximately equal to

0.995 and this means that only 0.5% of the emitter current consists of electrons passing from the base to the emitter.

Immediately the holes cross the emitter/base junction, and are said to have been emitted or injected into the base, they become minority charge carriers and start to diffuse across the base towards the collector/base junction. Because the base is fairly narrow and is also lightly doped, most of the emitted holes reach the collector/base junction and do not recombine with a free electron on the way. On reaching the junction, the emitted holes augment the minority charge carrier current crossing the junction and cause an increase in the collector current. The ratio of the number of holes arriving at the collector to the number of emitted holes is known as the *base transmission factor*, symbol β. Typically $\beta = 0.995$.

(1) The collector current is less than the emitter current because (a) part of the emitter current consists of electrons that do not contribute to the collector current and (b) not all of the holes injected into the base are successful in reaching the collector. Factor (a) is represented by the emitter injection ratio and factor (b) by the base transmission factor; hence the ratio of collector current to emitter current is equal to $\beta\gamma$. Substituting the typical values quoted for γ and β shows that, typically, the collector current is about 0.99 times the emitter current.

(2) The base current is small and has three components: (a) an electron current entering the base to replace the electrons lost by recombination with the diffusing holes, (b) the majority charge carrier electron current flowing from base to emitter, and (c) the collector leakage current I_{CBO}. The first two of these components are currents that flow out of the base and together are greater than I_{CBO} which flows into the base, and so the total base current flows out of the base. The total current flowing into the transistor must be equal to the total current flowing out of it and hence the emitter current I_e is equal to the sum of the collector and base currents, I_c and I_b respectively, that is

$$I_e = I_c + I_b \tag{3.1}$$

Typically, I_c is equal to $0.99I_e$ so that I_b is equal to $0.01I_e$.

(3) If the emitter current is varied by some means, the number of holes arriving at the collector, and hence the collector current, will vary accordingly. The magnitude of the collector-base voltage V_{cb} has relatively little effect on the collector current as will be seen shortly. Control of the output (collector) current can thus be obtained by means of the input

current to the emitter and this, in turn, can be controlled by variation of the bias voltage applied to the emitter/base junction. An increase in the bias voltage (which is in the forward direction) lowers the height of the potential barrier and allows an increased emitter current to flow; conversely, a decrease in the bias voltage reduces the emitter current.

(4) The ratio of the output current of a transistor to its input current in the absence of an a.c. signal is known as the D.C. CURRENT GAIN of the transistor. In the previous discussion the output current has been the collector current I_c and the input current has been the emitter current I_e. Thus,

$$\text{d.c. current gain,} - h_{FB} = \frac{I_c}{I_e} \qquad (3.2)$$

The minus sign indicates that the input and output currents are flowing in opposite directions. By convention, a current flowing into a transistor is taken to be positive and a current flowing out is taken to be negative. Since the operation of the transistor depends upon the movement of both holes and electrons the device should really be called the "bipolar transistor".

(5) A transistor may be connected in a circuit in one of three ways and in each case one electrode is common to both input and output. The connection is then described in terms of the common electrode; for example, the common-base connection has the base common to both input and output, the input signal is fed between the emitter and the base, and the output signal is developed between the collector and the base. In all connections, the base/emitter junction is always forward-biased and the collector/base junction is always reverse-biased.

The Common-base Connection

The basic arrangement of the common-base connection (or configuration) is shown in Fig. 3.4. The transistor has an alternating source of e.m.f. E_s volts r.m.s. and internal resistance R_s ohms connected to its input terminals. The alternating source is connected in series with the emitter/base voltage E_{eb} and varies the forward bias applied to the emitter/base junction.

During positive half-cycles of the source e.m.f., the forward bias applied to the junction is increased, the potential barrier is lowered, and an enhanced emitter current flows into the transistor. Conversely, during negative half-cycles the emitter current is decreased and in this way the collector current is caused to vary in accordance with the waveform of the alter-

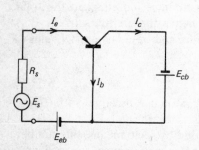

Fig. 3.4 The common-base connection

nating source. The collector/base bias battery E_{cb} has negligible internal resistance and so the collector/base voltage remains constant as the collector current varies. The collector circuit is said to be short-circuited so far as alternating currents are concerned.

In a common-base amplifier circuit an important parameter is the SHORT-CIRCUIT CURRENT GAIN of the transistor, symbol h_{fb}. The short-circuit current gain is defined as the ratio of a change in collector current to the change in emitter current producing it, with the collector/base voltage maintained constant, that is

$$\boxed{h_{fb} = \frac{\delta I_c}{\delta I_e} \quad \text{when } V_{cb} \text{ is constant}} \tag{3.3}$$

The *short-circuit* current gain is specified since analysis shows that the current gain is a function of the value of any resistance placed in the collector circuit. For the common-base circuit, however, the difference between the short-circuit current gain and the current gain for any particular collector load resistance is very small for all resistance values used in practical circuits and will be neglected in this book.

EXAMPLE 3.1

In a certain transistor a change in emitter current of 1 mA produces a change in collector current of 0.99 mA. Determine the short-circuit current gain of the transistor.

Solution

$$\text{Current gain } h_{fb} = \frac{\delta I_c}{\delta I_e} = \frac{0.99}{1} = 0.99 \quad (Ans.)$$

This is a typical value for the short-circuit current gain of a transistor connected in the common-base configuration. It should be evident that h_{fb} must be less than unity, because the emitter current is the sum of the base and collector currents. Clearly, then, a common-base transistor must have a current gain of less than unity but, if a resistor is connected in the collector circuit, as shown in Fig. 3.5, both voltage and power gains are possible.

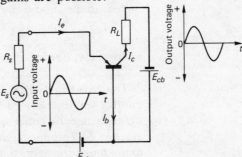

Fig. 3.5 The basic common-base amplifier

The output voltage is developed across the collector load resistor and, since the internal resistance of the collector supply is negligible, the top end of the resistor is effectively at earth potential so far as alternating currents are concerned. Thus the output signal voltage is taken from between the collector electrode and earth.

The source of the output power is the collector/base bias battery, the transistor effectively acting as a device for the conversion of d.c. power from the battery into the a.c. power supplied to the load.

In the common-base amplifier the input signal voltage and the output signal voltage are in phase, as shown by the waveforms of Fig. 3.5. Consider the input signal voltage to be passing through zero and increasing in the positive direction. The forward bias of the base/emitter junction is then increased and this results in an increase in the emitter current. The collector current is increased and the voltage drop across the collector load resistor R_L increases also and this makes the collector/base potential less negative. Thus a positive increment in the input signal voltage produces a positive increment in the output signal voltage.

The Common-emitter Connection

In practice, transistors are most often used in the common-emitter configuration shown in Fig. 3.6.

The emitter/base junction is forward-biased by the battery E_{be} and the collector/base junction is reverse-biased by a potential equal to $(E_{ce} - E_{be})$. However, since the voltage of the collector/emitter bias battery E_{ce} is much larger than the emitter/base bias voltage E_{be}, the reverse bias voltage may be taken as merely equal to E_{ce} volts.

When a transistor is connected in this way, the input current is the base current and not the emitter current as previously. During the negative half-cycles of the input signal voltage E_s, the forward bias of the emitter/base junction is increased, and so the emitter current I_e is increased by an amount δI_e. The collector current is also increased, by an amount $\delta I_c = h_{fb}\delta I_e$, and so is the base (input) current, by an amount

$$\delta I_b = \delta I_e - \delta I_c = \delta I_e (1 - h_{fb})$$

Conversely, during positive half-cycles of the input signal voltage the three currents are reduced in magnitude.

The SHORT-CIRCUIT CURRENT GAIN of a common-emitter connected transistor, symbol h_{fe}, is defined as the ratio of a change in collector current δI_c to the change in base current δI_b producing it, the collector/emitter voltage being maintained constant, that is

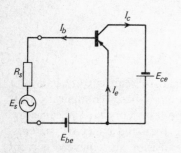

Fig. 3.6 The common-emitter connection

$$\boxed{h_{fe} = \frac{\delta I_c}{\delta I_b} \quad \text{when } V_{ce} \text{ is constant}} \qquad (3.4)$$

$$h_{fe} = \frac{h_{fb}\delta I_e}{\delta I_e - h_{fb}\delta I_e}$$

$$= \frac{h_{fb}\delta I_e}{\delta I_e(1 - h_{fb})}$$

$$= \frac{h_{fb}}{1 - h_{fb}} \qquad (3.5)$$

Typical values for the short-circuit current gain h_{fb} of a common-base transistor are in the neighbourhood of unity and thus the common-emitter connection can give a considerable current gain.

EXAMPLE 3.2

A transistor exhibits a change of 0.995 mA in its collector current for a change of 1 mA in its emitter current. Calculate (*a*) its common-base short-circuit current gain and (*b*) its common-emitter short-circuit current gain.

Solution
(*a*) Common-base short-circuit current gain

$$h_{fb} = \frac{\delta I_c}{\delta I_e} = \frac{0.995}{1} = 0.995 \quad (Ans.)$$

(*b*) Common-emitter short-circuit current gain

$$h_{fe} = \frac{\delta I_c}{\delta I_b} = \frac{h_{fb}}{1 - h_{fb}} = \frac{0.995}{1 - 0.995} = 199 \quad (Ans.)$$

When a load resistor R_L is connected in the collector circuit (Fig. 3.7) the current gain of the transistor is no longer equal to the short-circuit value but is somewhat less. The actual value of the current gain is dependent upon the value of the collector load resistor R_L, decreasing with increase in R_L.

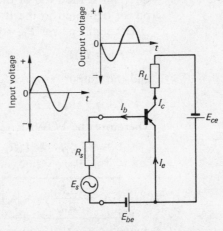

Fig. 3.7 The basic common-emitter amplifier

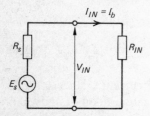

Fig. 3.8 Circuit for the calculation of the input current to a transistor

To obtain expressions for the voltage gain and the power gain of a transistor connected with common emitter: let the r.m.s. voltage and internal resistance of the voltage source applied to the input terminals of the transistor be E_s volts and R_s ohms respectively, and let the input resistance of the transistor be R_{IN} ohms.

Then (Fig. 3.8), the input current to the transistor, I_b, is

$$I_b = \frac{E_s}{R_s + R_{IN}}$$

and the voltage V_{IN} appearing across the transistor input terminals is

$$V_{IN} = I_b R_{IN} \quad \text{or} \quad I_b = \frac{V_{IN}}{R_{IN}}$$

The output or collector current I_c is

$$I_c = \frac{h_{fe} V_{IN}}{R_{IN}}$$

This current flows through the collector load resistance R_L and develops the output voltage V_{OUT} across it, therefore

$$V_{OUT} = \frac{h_{fe} V_{IN}}{R_{IN}} R_L$$

and the VOLTAGE GAIN A_v is

$$\boxed{A_v = \frac{V_{OUT}}{V_{IN}} = \frac{h_{fe} R_L}{R_{IN}}} \tag{3.6}$$

Since the short-circuit current gain h_{fe} is greater than unity and the collector load resistance R_L is usually greater than the input resistance R_{IN} of the transistor, a voltage gain is readily achieved.

Now consider the power gain of a common-emitter transistor. This is the ratio of the power delivered to the load to the power delivered to the transistor.

The input power P_{IN} to the transistor is (see Fig. 3.8)

$$P_{IN} = I_b^2 R_{IN}$$

and the output power P_{OUT} is

$$P_{OUT} = (h_{fe} I_b)^2 R_L$$

Therefore the POWER GAIN A_p is

$$A_p = \frac{P_{OUT}}{P_{IN}} = \frac{h_{fe}^2 I_b^2 R_L}{I_b^2 R_{IN}}$$

$$\boxed{A_p = \frac{h_{fe}^2 R_L}{R_{IN}} = A_v A_i} \tag{3.7}$$

Again, since R_L is greater than R_{IN} a power gain is possible.

EXAMPLE 3.3

A transistor is connected with common-emitter in a circuit and has a collector load resistance of 2000 Ω. The short-circuit current gain of the transistor is 100 and its input resistance is 1000 Ω. Calculate the voltage and power gains of the transistor.

Solution
From equation (3.6),

$$\text{Voltage gain} = \frac{h_{fe}R_L}{R_{IN}} = \frac{100 \times 2000}{1000} = 200 \qquad (Ans.)$$

From equation (3.7),

$$\text{Power gain} = \frac{h_{fe}^2 R_L}{R_{IN}} = 200 \times 100 = 20\,000 \qquad (Ans.)$$

The Common-collector Connection

The third way in which a transistor may be connected is shown in Fig. 3.9. The collector is now common to both input and output circuits and the load is connected in the emitter circuit. With this configuration the base current is the input current and the emitter current is the output current. The SHORT-CIRCUIT CURRENT GAIN is defined as

$$\text{Short-circuit current gain} = \frac{\delta I_e}{\delta I_b} \text{ when } V_{ce} \text{ is constant}$$

$$(3.8)$$

$$= \frac{\delta I_e}{\delta I_e - \delta I_c}$$

$$= \frac{\delta I_e}{\delta I_e(1 - h_{fb})} \qquad (3.9)$$

$$= \frac{1}{1 - h_{fb}} \qquad (3.10)$$

$$= h_{fe} + 1 \qquad (3.11)$$

The short-circuit current gain of a transistor connected in the common-collector configuration is approximately equal to the short-circuit gain of the same transistor connected with common-emitter. The current gain when a load is connected in the emitter circuit is not equal to the short-circuit current gain but is reduced by an amount that is dependent on the value of the emitter load.

Expressions (3.6) and (3.7) may be used to determine the voltage and power gains of a common-collector circuit. Now, however, the input resistance is considerably larger than the load resistance and a voltage gain of less than unity is obtained.

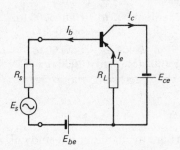

Fig. 3.9 The basic common-collector amplifier

The main use of the common-collector circuit—or the "emitter-follower" as it is often called—is as a power-amplifying impedance transformer that can be connected between a high impedance source and a low impedance load.

A comparison between the main characteristics of the three transistor configurations is given in Table 3.1.

Table 3.1

Characteristic	Common-base	Common-emitter	Common-collector
Short-circuit current gain	h_{fb}	$\dfrac{h_{fb}}{1-h_{fb}}$	$\dfrac{1}{1-h_{fb}}$
Voltage gain	Good	Better than common-base	Unity or less
Input-resistance	Low 30 to 100 Ω	Medium 800 to 5000 Ω	High 5000 to 500 000 Ω
Output-resistance	High 10^5 to 10^6 Ω	High 10 000 to 50 000 Ω	Low 50 to 1000 Ω

Transistor Static Characteristics

A number of current/voltage plots are available in the study of the operation of a transistor in a circuit. The resulting curves, which are known as the static characteristic curves, give information on the value of current flowing into or out of one electrode for either a given current flowing into or out of another electrode or a given voltage applied between two electrodes. Four sets of characteristics can be plotted for each configuration: (a) the input characteristic, (b) the transfer characteristic, (c) the output characteristic, and (d) the mutual characteristic. In this book, however, the characteristics for the common-collector circuit will not be discussed.

Common-base Static Characteristics

The method of determining the static characteristics of a transistor is to connect the transistor into a suitable circuit and then to vary the appropriate currents and/or voltages in a number of discrete steps, noting the corresponding values of other currents at each step. Fig. 3.10 shows a suitable circuit for the determination of the characteristics of a p-n-p transistor in the common-base configuration.

The collector and base currents are shown as flowing *out* of the transistor and are therefore, by definition, negative; the emitter current is shown flowing *into* the transistor and must

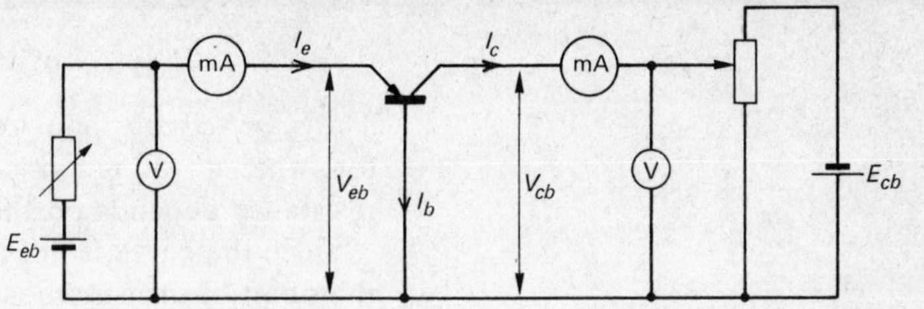

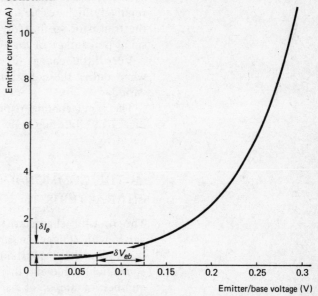

Fig. 3.10 Circuit for the determination of the static characteristics of a common-base connected transistor

be taken as positive. If the characteristics of an n-p-n transistor had to be measured, the polarities of the two batteries would have to be reversed.

(*a*) THE COMMON-BASE INPUT CHARACTERISTIC

The input characteristic of a transistor connected with common base shows how the emitter current varies with change in the emitter/base voltage, with the collector/base voltage held constant.

Fig. 3.11 Common-base input characteristic
$\delta V_{eb} = 0.125 - 0.075 = 0.05\ \text{V}$;
$\delta I_e = 0.9 - 0.4 = 0.5\ \text{mA}$

The method of determining the input characteristic is as follows. The collector/base voltage is adjusted to a convenient value and then the emitter/base voltage is increased in a number of discrete steps, the resulting value of emitter current being noted at each step. The values of emitter current obtained in this way are then plotted against the corresponding values of emitter/base voltage. Fig. 3.11 shows a typical characteristic. The slope of the characteristic, that is $\delta I_e/\delta V_{eb}$, is the input conductance of the transistor with its output terminals short-circuited to alternating current, and the short-circuit input resistance R_{IN} is given by the reciprocal of the slope:

$$R_{IN} = \frac{\delta V_{eb}}{\delta I_e} \quad (V_{cb} \text{ constant}) \tag{3.12}$$

The short-circuit input resistance is a property of the transistor: when a load is connected to the output terminals, the input resistance depends on the load resistance.

Since the curve is not linear, the value of R_{IN} will vary with the point of measurement. The input resistance R_{IN}, when the emitter/base voltage V_{eb} is 0.1 V, is from Fig. 3.11,

$$R_{IN} = \frac{0.05}{0.5 \times 10^{-3}} = 100 \ \Omega$$

It is evident that at the point $V_{eb} = 0.25$ V the input resistance will be less than 100 ohms because at this point a change in V_{eb} of 0.05 V gives a change in emitter current in excess of 0.5 mA. The change in the input resistance with change in the emitter/base voltage gives rise to distortion of signals handled by the transistor. The effect can be minimized by connecting a relatively high resistance in series with the input terminals of the transistor so that any change in input resistance is only a small percentage of the total resistance in the input circuit.

Very little change in the input characteristic takes place when either the collector/base voltage or the temperature is varied.

The input characteristic of an n-p-n transistor is similar to Fig. 3.11, differing only in that both I_e and V_{be} would be negative.

(b) THE COMMON-BASE CURRENT-TRANSFER CHARACTERISTIC

The transfer characteristic shows how the collector current varies with changes in emitter current, the collector/base voltage being held constant. The collector voltage is set to a convenient value and then the emitter current is increased in a number of steps; at each step the corresponding value of collector current is noted. A typical transfer characteristic is to be seen in Fig. 3.12 and is more or less independent of changes in either the collector/base voltage or the temperature. For an n–p–n transistor I_c would be positive and I_e negative.

The slope of the transfer characteristic gives the short-circuit current gain h_{fb} of the transistor. At the point $I_e = 5$ mA,

$$h_{fb} = \frac{\delta I_c}{\delta I_e} \quad (V_{cb} \text{ constant})$$

$$= 1.9/2 = 0.95$$

Although the short-circuit current gain of a transistor can be determined from its transfer characteristic, the transfer characteristic is not often used because the same information can be obtained from the output characteristic.

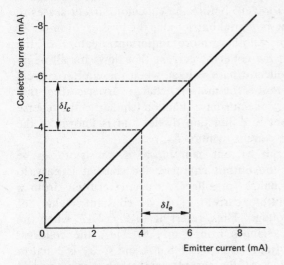

Fig. 3.12 Common-base current-transfer characteristic
$\delta I_c = -5.8 - (-3.9) = -1.9\,\text{mA}$;
$\delta I_e = 6 - 4 = 2\,\text{mA}$

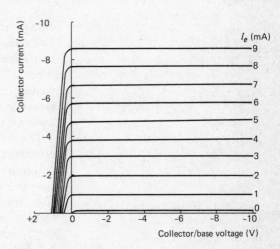

Fig. 3.13 Common-base output characteristics

(c) THE COMMON-BASE OUTPUT CHARACTERISTIC

The output characteristic indicates the way in which the collector current varies with change in collector/base voltage, with the emitter current held constant. The emitter current is set to a convenient low value and the collector/base voltage is increased from zero in a number of discrete steps and at each step the collector current that flows is noted. The collector/base voltage is then restored to zero and the emitter current is increased to another convenient value and the procedure repeated. In this way a whole family of curves relating collector current to collector/base voltage can be obtained, a typical family being shown in Fig. 3.13. The output characteristics of a n-p-n transistor are of similar shape but both I_c and V_{ce} would be positive.

The reciprocal of the slope of the output characteristic, $\delta V_{cb}/\delta I_c$, gives the output resistance of the transistor, when the input terminals are open-circuited to alternating current, at the point where the measurement is made. The open-circuit output resistance is a property of the transistor: when a signal is applied to the input terminals, the output resistance depends on the resistance of the signal source. Since the curves are linear over most of their length, the output resistance is fairly constant, and since the curves are nearly parallel to the collector/base voltage axis the output resistance is very high, of the order of $100 \text{ k}\Omega$ or more.

It can be seen that some collector current still flows when the collector/base voltage has been reduced to zero. This is because the potential barrier across the collector/base junction must be reduced to zero before the collector current ceases to flow (because the potential barrier aids the passage of minority charge carriers). Another, more important, feature of the characteristics is the collector current that flows for all negative values of collector/base voltage when the emitter current is zero. This current is the minority charge carrier current that passes across the collector/base junction (similar to the reverse saturation current in a junction diode) and is known as the *collector leakage current*, symbol I_{CBO}.

The short-circuit current gain h_{fb} of a transistor can be estimated from the output characteristic since it is easy to determine the change in collector current resulting from a change in emitter current, for a constant value of collector/base voltage. Thus, referring to Fig. 3.13, when the emitter current changes from 5 mA to 7 mA the collector current changes from 4.9 mA to 6.8 mA and so h_{fb} is equal to 1.9/2 or 0.95. Because the current gain is less than unity and the input and output impedances are so different the common-base connection is rarely employed for audio-frequency circuits.

Common-emitter Static Characteristics

To determine the static characteristics of a transistor connected in the common-emitter configuration, the transistor must be connected, with common emitter, in a circuit similar to that given in Fig. 3.10; the only circuit alterations necessary are the removal of the milliammeter from the emitter circuit and the insertion of a microammeter in the base circuit.

(a) COMMON-EMITTER INPUT CHARACTERISTIC

The input characteristic shows the way in which the base current varies with change in the base/emitter voltage, the collector/emitter voltage remaining constant. The method of determining the input characteristic is to maintain the collector/emitter voltage constant at a convenient value and increase the base/emitter voltage in a number of discrete steps, noting the base current at each step. The procedure is then repeated for a different but constant value of collector/emitter voltage V_{ce}, since change in this voltage has an effect on the input characteristic. A typical pair of input characteristics are shown in Fig. 3.14. The input resistance, for a given

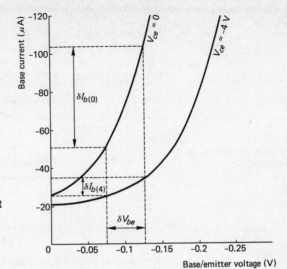

Fig. 3.14 Common-emitter input characteristic
$\delta V_{be} = 0.125 - 0.075 = 0.05\,V$;
$\delta I_{b(0)} = 104 - 51 = 53\,\mu A$;
$\delta I_{b(4)} = 35 - 26 = 9\,\mu A$

base/emitter voltage V_{be} is given by the reciprocal of the slope of the curve at that point. For example, consider the short-circuit input resistance of the transistor at the point $V_{be} = 0.1\,V$ for both $V_{ce} = 0$ and $V_{ce} = -4\,V$.

For $V_{ce} = 0\,V$

$$R_{IN} = \frac{\delta V_{be}}{\delta I_b} \qquad (V_{ce}\ constant) \qquad (3.13)$$

$$= \frac{0.05}{53 \times 10^{-6}} = 943\,\Omega$$

and for $V_{ce} = -4\,V$

$$R_{IN} = \frac{0.05}{9 \times 10^{-6}} = 5556\,\Omega$$

The corresponding n-p-n characteristics would have positive values of I_b, V_{be} and V_{ce}.

(b) COMMON-EMITTER CURRENT-TRANSFER CHARACTERISTIC

The transfer characteristic shows how the collector current changes with changes in the base current, the collector/emitter voltage being held at a constant value. For this measurement the collector/emitter voltage is kept constant and the base current is increased in a number of discrete steps and at each step the collector current is noted. Finally a plot is made of collector current against base current. Since the transfer characteristic is not independent of the value of the collector/emitter voltage, the procedure is repeated for a number of different collector/emitter voltages to give a family

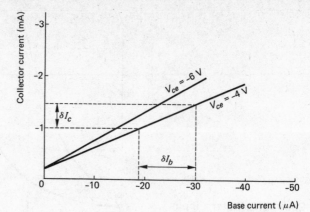

Fig. 3.15 Common-emitter current-transfer characteristic
$\delta I_c = -1.5 - (-1) = -0.5\,\text{mA};$
$\delta I_b = -30 - (-19) = -11\,\mu\text{A}$

of curves, Fig. 3.15 showing a typical p-n-p transistor characteristic.

The slope of the transfer characteristic gives the short-circuit current gain h_{fe} of the transistor when it is connected in the common-emitter configuration. Thus for the curve marked $V_{ce} = -4\,\text{V}$,

$$h_{fe} = \frac{\delta I_c}{\delta I_b} = \frac{0.5 \times 10^{-3}}{11 \times 10^{-6}} = 45.45$$

(c) COMMON-EMITTER OUTPUT CHARACTERISTIC

The output characteristic illustrates the changes that occur in collector current with changes in collector/emitter voltage, for constant value of base current. The base current is set to a convenient value and is maintained constant and the collector/emitter voltage is increased from zero in a number of discrete steps, the collector current being noted at each step. The collector/emitter voltage is then restored to zero and the base current increased to another convenient value and the procedure repeated. In this way a family of curves (see Fig. 3.16) can be obtained. For the corresponding n-p-n characteristic the polarities of I_c, I_b and V_{ce} should be changed to positive.

The open-circuit output resistance of the transistor is equal to the reciprocal of the slope of the output characteristic. The open-circuit output resistance at the point $V_{ce} = -6\,\text{V}$ and $I_b = -60\,\mu\text{A}$ is

$$R_{OUT} = \frac{\delta V_{ce}}{\delta I_c} \qquad (I_b \text{ constant}) \tag{3.14}$$

$$= \frac{2}{0.2 \times 10^{-3}} = 10\,000\,\Omega$$

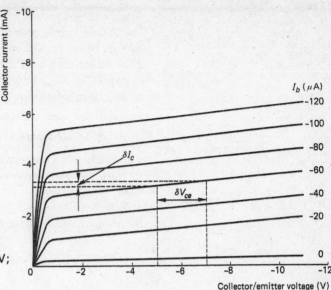

Fig. 3.16 Common-emitter output characteristics $\delta V_{ce} = -7 - (-5) = -2\,\text{V}$; $\delta I_c = -3.3 - (-3.1) = -0.2\,\text{mA}$

When a characteristic is non-linear its slope will vary according to the point of measurement and therefore the point of measurement should always be quoted. It is usual, unless specified otherwise, to measure the slope of the most linear portion of a characteristic. For the greatest accuracy the increments taken either side of the chosen point should be as small as possible although this has not been done in this chapter in order to clarify the diagrams.

The output characteristics can also be used to determine the short-circuit current gain h_{fe} of the transistor, since, for a given value of collector/emitter voltage V_{ce}, the change in collector current δI_c produced by a change in base current δI_b can be obtained by projecting from the appropriate curves. Thus, for $V_{ce} = -4\,\text{V}$ a change in the base current from $-40\,\mu\text{A}$ to $-60\,\mu\text{A}$ will produce a change in the collector current from 2.1 to 2.9 mA. The current gain h_{fe} is therefore equal to

$$\frac{(2.9 - 2.1) \times 10^{-3}}{(60 - 40) \times 10^{-6}} = 40$$

It will be seen from Fig. 3.16 that a collector current flows even when the input or base current is zero. This current is the *common-emitter leakage current*, symbol I_{CEO}, which is related to the common-base leakage current I_{CBO} according to the expression

$$I_{CEO} = I_{CBO}(1 + h_{fe}) \tag{3.15}$$

The collector leakage current I_{CBO} of a common-base transistor is extremely temperature-sensitive and is approximately doubled for every 12 degC rise in temperature for silicon transistors and every 8 degC rise for germanium transistors. However, the leakage current of a silicon transistor at a given temperature is much less than the leakage current of an equivalent germanium transistor at the same temperature.

Typically, I_{CBO} at 20°C for a germanium transistor may be about 10 μA but only about 50 nA for a silicon transistor.

(d) COMMON-EMITTER MUTUAL CHARACTERISTICS

The mutual characteristics of a common-emitter connected transistor show the changes in collector current that occur with changes in the base/emitter voltage, with the collector/emitter voltage held constant. The slope of the mutual characteristic is the mutual conductance of the transistor. Thus

$$g_m = \frac{\delta I_c}{\delta V_{be}} \tag{3.16}$$

$$= \frac{\delta I_c}{\delta I_b} \times \frac{\delta I_b}{\delta V_{be}}$$

$$= \frac{h_{fe}}{R_{IN}} \tag{3.17}$$

For all transistors the mutual conductance is approximately equal to 40 mS/mA of collector current. Thus, if the peak a.c. collector current is, say, 2 mA, the mutual conductance g_m would be 80 mS.

The Construction of Transistors

A number of different methods of manufacturing transistors have been developed since the invention of the transistor in 1948 and a description of many of them is outside the scope of this book. The most commonly employed types of transistor are the alloy junction transistor and the silicon planar transistor and only the construction of these types will be presented.

The construction of a germanium alloy junction transistor is shown in Fig. 3.17; the method is an extension of the method previously described for the alloy junction diode. The construction of a SILICON PLANAR TRANSISTOR is shown in Fig. 3.18. The steps involved in the manufacture of a silicon planar transistor are as follows: a wafer of n-type silicon is oxidized to a depth of approximately 1 micron (Fig. 3.19a) and then the oxide is partially etched off (Fig. 3.19b). Next the wafer is exposed to a vapour of the acceptor element boron and the impurity is allowed to diffuse into the wafer to a predetermined depth. At the same time the wafer surface is reoxidized (Fig. 3.19c). A part of the reoxidized surface is

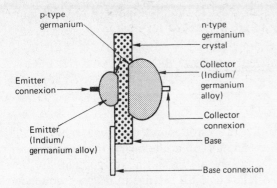

Fig. 3.17 Construction of an alloy junction transistor

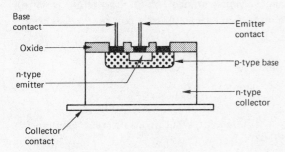

Fig. 3.18 Construction of a silicon planar transistor

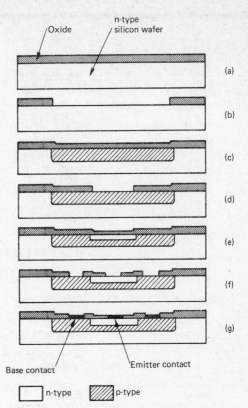

☐ n-type ▨ p-type

Fig. 3.19 The stages in the manufacture of a silicon planar transistor

then etched away (Fig. 3.19*d*) and the wafer exposed to a vapour of the donor element phosphorus and is also reoxidized again (Fig. 3.19*e*). The wafer now contains a layer of p-type material that will form the base of the transistor and a layer of n-type material that provides the emitter. The wafer is then etched to separate the base and emitter regions on the surface of the wafer (Fig. 3.19*f*) and finally (Fig. 3.19*g*) metal contacts are alloyed onto the etched areas.

The wafer is cut to the required size, mounted on a suitable collector contact, and then leads are connected to the base and emitter contacts.

The construction of a planar transistor requires a relatively thick collector wafer in order to give adequate mechanical support to the other layers and this increases the resistance of the collector. For some applications the collector resistance is too large and must be reduced. If the resistance were to be reduced by the use of a low-resistivity material for the collector, the breakdown voltage of the transistor would also be reduced and the collector/base capacitance would be increased, both undesirable effects. To overcome these effects an epitaxial layer is employed in the collector. This is a layer, approximately 0.1 mm thick, of high-resistivity material that is deposited on the main collector and that permits the collector to be made from a material of low resistivity.

Thermal and Frequency Effects

Frequency Characteristics

The current gain of a transistor is not the same value at all frequencies, but, instead, falls off at the higher frequencies (see Fig. 3.20). The frequency at which the magnitude of the current gain h_{fe} has fallen by 3 dB relative to its low-frequency value h_{feo} is known as the CUT-OFF FREQUENCY f_β of the transistor. Eventually, at some frequency f_1, $|h_{fe}|$ falls to unity. The high-frequency performance of a transistor is often quoted by the manufacturer in terms of a parameter f_t, where

$$f_t = |h_{fe}| \cdot f \tag{3.18}$$

f being any frequency.

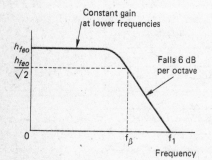

Fig. 3.20 Variation of $|h_{fe}|$ with frequency

EXAMPLE 3.4

A transistor has $f_t = 500\,\text{MHz}$. What is its current gain at (a) 100 MHz, (b) 10 MHz?

Solution
(a) $f_t = 500\,\text{MHz} = |h_{fe}| \times 100\,\text{MHz}$

   $|h_{fe}| = 5$ (*Ans.*)

(b) $f_t = 500\,\text{MHz} = |h_{fe}| \times 10\,\text{MHz}$

   $|h_{fe}| = 50$ (*Ans.*)

Thermal Runaway

An increase in the temperature of the collector/base junction will cause the collector leakage current I_{CBO} to increase. The increase in collector current produces an increase in the power dissipated at the junction and this, in turn, further increases the temperature of the junction and so gives further increase in I_{CBO}. The process is cumulative and, particularly in the common-emitter connection ($I_{CEO} \gg I_{CBO}$), may lead to the eventual destruction of the transistor. In practice, thermal runaway is prevented in well-designed circuits by the use of stabilization circuitry that compensates for any increase in I_{CBO} and also, for power transistors, by the use of a heat sink to provide rapid conduction of heat away from the junction.

The manufacturer of a transistor quotes the maximum permissible power that can be dissipated within the transistor without damage. The power dissipated within a transistor is predominantly the power which is dissipated at its collector-

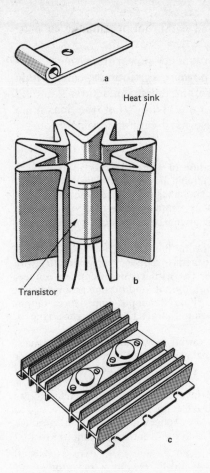

Fig. 3.21 Three kinds of heat sink

base junction, and this, in turn, is equal to the d.c. power taken from the collector supply voltage minus the total output power (d.c. power plus a.c. power). For transistors handling small signals, the power dissipated at the collector is small and there is generally little problem. When the power dissipated by the transistor is large enough to cause the junction temperature to rise to a dangerous level, it is necessary to improve the rate at which heat is removed from the device. Power transistors are constructed with their collector terminal connected to their metallic case. To increase the area from which the heat is removed the case of the transistor can be bolted on to a sheet of metal known as a HEAT SINK. Heat will move from the transistor to the heat sink by conduction and be removed from the sink by convection and radiation.

The simplest heat sink is shown in Fig. 3.21a. It consists of a push-fit clip that can be screwed on to a metal chassis, the transistor being a push-fit into the hole. The greater the power dissipated within a transistor, the larger the surface area of the heat sink required to remove sufficient heat to keep the junction temperature with safe limits. To prevent the heat sink occupying too much space within equipment, it is common to employ structures of the kinds shown in Figs. 3.21b and 3.21c.

For maximum efficiency a heat sink should (i) be in excellent thermal contact with the transistor case, (ii) have the largest possible surface area and be painted matt black, and (iii) be mounted in a position such that a free flow of air past it is possible.

Transistor Ratings

The parameters of major concern in selecting a transistor type for a particular circuit application are

(1) The maximum allowable collector/base voltage.
(2) The maximum power dissipation in the transistor at 25°C.
(3) The current gain h_{fe}.
(4) The f_t of the transistor.

Generally, one or other of these four parameters is of prime importance and will be the deciding factor in selecting a transistor.

For example, if a transistor is required for a small-signal audio-frequency amplifier, the only parameters of importance will be the current gain and perhaps the collector/base voltage. This is because the powers involved will be small enough to be handled easily by any transistor and high-frequency operation is not required. In most cases the collector/base voltage is unlikely to be an important factor since the voltages in the amplifier will probably be well within the capabilities of most a.f. transistors.

When choosing an audio frequency power transistor, care is necessary to ensure that the power dissipation expected within the transistor will be well within the manufacturer's quoted maximum value, and it is quite likely that the maximum collector/base voltage will also need careful consideration.

Exercises

3.1. Describe briefly the operation of a transistor as an amplifying device. With the aid of a circuit diagram show how you would determine the basic static characteristics of a transistor.

Sketch the current/voltage characteristics that would result from such measurements, giving typical values on your axes.

(C & G)

3.2. Draw circuit diagrams showing how a junction transistor may be used in a single-stage audio-frequency amplifier (*a*) with common-base connection, (*b*) with common-emitter connection. Mark clearly the input, output, and biasing arrangements, and the current flowing. State the input impedance, output impedance, and current gain characteristics for each type of amplifier.

Derive an expression showing how, for small input signals, the current gain of a transistor connected in a common-emitter circuit is related to the gain of the same transistor in a common-base circuit.

A transistor connected in a common-emitter circuit shows changes in emitter and collector currents of 1.0 mA and 0.98 mA, respectively. What changes in base current produce these changes and what is the current gain of the transistor?

(C & G)

3.3. The data given in Tables A and B refer to a transistor in the common-emitter configuration.

Table A

Collector voltage V_{ce} (volts)		−2	−4	−6	−8	−10
Collector current (mA)	Base current −120 μA	−6.4	−7	−7.6	−8.2	−8.8
	Base current −80 μA	−4.4	−4.8	−5.2	−5.6	−6.0
	Base current −40 μA	−2.2	−2.4	−2.6	−2.8	−3.0

Use the data in Table A to plot the collector-voltage/collector-current characteristics for $I_b = -40$, -80 and -120 μA, and from the characteristic for $I_b = -80$ μA deduce the output resistance of the transistor. Use the data in Table B to plot the collector-current/base-current characteristic for $V_{ce} = -4.5$ V and from the graph deduce the current gain.

(C & G)

Table B

Collector voltage $V_{ce} = -4.5$ volts				
Base current $I_b (\mu A)$	0	−5	−10	−15	−20
Collector current I_c (mA)	−0.15	−0.45	−0.75	−1.1	−1.4

3.4. Explain the principle of the transistor.
Why is it necessary to avoid excessive temperature rise in the transistors of an amplifier? (C&G)

3.5. The data given in Table C refers to a transistor in the common-emitter configuration.

Table C

Collector/emitter voltage (V)	Collector current (mA)		
	Base current −60 μA	Base current −90 μA	Base current −120 μA
−1	−3.1	−4.6	−6.0
−3	−3.5	−5.1	−6.6
−5	−3.9	−5.6	−7.2
−7	−4.3	−6.1	−7.8
−9	−4.7	−6.6	−8.4

Use the data to plot the collector current/collector voltage characteristics for $I_b = -60$, −90 and −120 μA. Use these characteristics to derive: (a) the output resistance of the transistor for $I_b = -90 \mu A$, (b) the current gain for $V_{ce} = -6$ V. Also, determine the current gain of the transistor when connected in a common-base circuit. (C&G)

3.6. The data given in Table D refer to a transistor in the common-emitter configuration.

Table D

Collector/emitter voltage (V)	Collector current (mA)		
	Base current −40 μA	Base current −60 μA	Base current −80 μA
−2	−2.9	−4.4	−5.9
−4	−3.3	−4.9	−6.4
−6	−3.7	−5.4	−7.0
−8	−4.1	−5.9	−7.6
−10	−4.5	−6.4	−8.2

Draw the collector current/collector voltage characteristics for $I_b = -40$, −60, −80 μA. Use these characteristics to determine: (a) the output resistance of the transistor for $I_b = -60 \mu A$, (b) the current gain for $V_{ce} = -7$ V. (C&G)

3.7. (*a*) Describe an experiment to determine the collector voltage/collector current characteristics for a transistor connected in the common-emitter configuration. Show how the current gain and output resistance can be deduced from these curves. (*b*) Table E gives values of the collector current/collector voltage for a series of base current values in a transistor in the common-emitter configuration. Plot these characteristics and hence find (i) the current gain when the collector voltage is 6 V, (ii) the output resistance for a base current of 45 μA.

Table E

Collector/emitter voltage (V)	Collector current (mA)			
	Base current 25 μA	Base current 45 μA	Base current 65 μA	Base current 85 μA
3	0.91	1.59	2.25	3.00
5	0.92	1.69	2.45	3.20
7	0.96	1.84	2.65	3.50
9	0.99	2.04	2.95	4.00

3.8. Use the input characteristics given in Fig. 3.14 and the output characteristics given in Fig. 3.16 to obtain values of collector current against corresponding values of base-emitter voltage for a collector/emitter voltage of -4 V. Use these values to plot the mutual characteristics of the transistor.

3.9. The data given in Table F refer to a transistor in the common-base configuration.

Table F

Collector/emitter voltage (V)	Collector current (mA)				
	Emitter current 0 mA	Emitter current 2 mA	Emitter current 4 mA	Emitter current 6 mA	Emitter current 8 mA
-5	0	-1.9	-3.7	-5.7	-7.6
-30	-0.1	-2.0	-3.8	-5.8	-7.7
-55	-0.2	-2.1	-3.9	-5.9	-7.8

Draw the collector current/collector voltage characteristics for the various values of emitter current and the collector current/emitter current curves for $V_{ce} = -30$ V. Calculate the current gain and output resistance of the transistor. (C&G)

3.10. A transistor has the data of Table G and is used in a common-emitter amplifier.

Plot the output characteristics assuming them to be linear between the values indicated. The collector supply voltage is 9 V and the collector load resistance is 3.8 kΩ. Draw the load line and use it to determine (i) the base bias current required to produce a quiescent collector current of 1 mA, (ii) the output voltage when an input base signal of 10 μA peak value is applied and (iii) the current gain.

Table G

Collector/emitter voltage (V)	Collector current (mA)			
	Base current 10 μA	Base current 20 μA	Base current 30 μA	Base current 40 μA
0.5	0.45	0.95	1.40	1.90
10	0.50	1.05	1.55	2.10

3.11. A common-emitter transistor has the output characteristics represented by Table H.

Table H

Collector/emitter voltage (V)	Collector current (A)		
	Base current 0 mA	Base current 10 mA	Base current 20 mA
1	—	0.64	1.28
2	0.04	—	—
8	—	—	1.40
12	—	0.72	—
16	0.10	—	—

Draw the characteristics and draw the load line for a collector supply voltage of 12 V and a collector load resistance of 10 Ω. Calculate the a.c. power dissipated in the load resistance when an input signal of 10 mA peak value is applied to the base. Mark clearly the operating point you have selected.

3.12. (a) Draw the circuit diagram of a simple resistance-loaded audio-frequency amplifier using a transistor in the common-emitter configuration. Show clearly the input, the output and the directions of current flow.

(b) Briefly explain how such an amplifier may be used to provide (i) current gain and (ii) voltage gain.

(c) In such an amplifier the current gain of the transistor is 49, the input resistance of the transistor is 1 kΩ and the collector load resistance 2 kΩ. Determine the voltage and power gains of the amplifier. (C&G)

3.13. (a) The data given in Table I refer to the output characteristics of a transistor.

Table I

Collector/emitter voltage (V)	Collector current (mA)		
	Base current 40 μA	Base current 60 μA	Base current 80 μA
1	3	4.5	6.0
3	3.4	5.0	6.5
5	3.8	5.5	7.0
9	4.2	6.0	7.6
11	4.6	6.5	8.2

(i) State what type of transistor this is and in which configuration it is connected.

(ii) Estimate the saturation voltage.

(b) Plot the output characteristics and determine (i) the output resistance of the transistor when $I_b = 60\ \mu A$, (ii) the current gain for $V_{ce} = 6\ V$, (iii) the mutual conductance when $I_b = 60\ \mu A$ and $V_{ce} = 6\ V$.

(c) The transistor is used in an amplifier having a collector supply voltage of 10 V and a load resistance of 1200 Ω. Draw the load line and choose a suitable operating point. Use your load line to calculate the current gain when a 20 μA peak signal is applied to the base.

Short Exercises

3.14. A transistor has a short-circuit current gain, when connected with common emitter, of 150. Calculate its short-circuit current gain when (i) in the common-base connection and (ii) in the common-collector connection.

3.15. Draw a circuit diagram of an arrangement suitable for the determination of the static characteristics of an n-p-n transistor connected with common emitter. Describe how you would use this circuit to obtain (i) the output characteristics and (ii) the mutual characteristics.

3.16. Compare the relative values of input and output resistance for the common-base, common-emitter, and common-collector connections. Quote typical values.

3.17. Sketch the basic arrangement of an n-p-n transistor connected as (i) a common base amplifier, (ii) a common-emitter amplifier, and (iii) a common-collector amplifier. For each circuit, deduce the phase relationship between an input sinusoidal voltage and the voltage appearing across the load resistor R_L.

3.18. What is meant by thermal runaway in a transistor?

3.19. Sketch a typical set of output characteristics for an n-p-n transistor.

3.20. Prove that the short-circuit current gain h_{fe} of a transistor connected with common-emitter is related to the common-base short-circuit current gain h_{fb} by the expression

$$h_{fe} = \frac{h_{fb}}{1 - h_{fb}}$$

3.21. A transistor has a current gain of 100 and input resistance of 1200 Ω, and is connected in an amplifier with a load resistor of 2200 Ω. Calculate the voltage and power gains obtained.

3.22. A transistor has a short circuit current gain of 0.992 and a collector leakage current of 12 μA when connected in the common-base configuration. What will be its collector leakage current when connected with common-emitter?

3.23. What is a heat sink? Draw a typical heat sink and list the factors which determine its efficiency.

3.24. What is meant by the shapes of typical input and transfer characteristics? Deduce and sketch the shape of a mutual characteristic. What information would be given by the slope of the characteristic?

4 Field-Effect Transistors

Introduction

The field-effect transistor (fet) is a semiconductor device which can perform many of the functions of a bipolar transistor, but which operates in a fundamentally different way. There are two kinds of fet available; the junction field-effect transistor (jfet), and the insulated gate field-effect transistor (igfet). The igfet is often known as the metal-oxide-semiconductor field-effect transistor (mosfet). This latter term will be used throughout this book. The mosfet can be sub-divided into two classes: the enhancement type and the depletion type. All three classes of fet can be obtained in either n-channel or p-channel versions and so a total of six different types of fet are available.

The fet combines some of the advantages of the bipolar transistor with a very high input impedance but it provides a smaller voltage gain.

The Junction Field-Effect Transistor

Operation

Fig. 4.1 shows a wafer of lightly-doped n-type silicon, provided with an ohmic contact at each of its two ends, and a battery applied between these contacts. The contact to which the positive terminal of the battery is connected is known as the DRAIN, whilst the negative side of the battery is connected to the SOURCE contact.

A current, consisting of majority charge carriers, will flow in the silicon wafer from drain to source, the magnitude of which is inversely proportional to the resistance of the wafer. This current is known as the DRAIN CURRENT. The resistance of the wafer in turn depends upon the resistivity of the n-type silicon

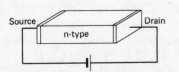

Fig. 4.1   n-type semiconductor

wafer and the length and cross-sectional area of the conduction path, or *channel*, i.e. $R = \rho l / a$. For given values of resistivity and length, the channel resistance will depend upon the cross-sectional area of the channel. If, therefore, the cross-sectional area can be varied by some means, the channel resistance and hence the drain current can also be varied.

The properties of a p-n junction are such that the region either side of the junction, known as the "depletion layer," is a region of high resistivity whose width is a function of the reverse-biased voltage applied to the junction. The depletion layer can be used to effect the required control of the channel resistance. A p-n junction is therefore required in the silicon wafer and to obtain one it is necessary to diffuse a p-type region into the wafer, as shown by Fig. 4.2. The p-type region is doped more heavily than the n-type channel to ensure that the depletion layer will lie mainly within the channel. An ohmic contact is provided to the p-type region and it is known as the gate terminal.

If the gate is connected directly to the source, the p-n junction will be reverse biased and the depletion layer will be extended further into the channel. The p-n junction is reverse biased because the p-type gate region is at zero potential, while the n-type channel region is at some positive potential. A potential gradient will exist along the length of the channel, varying from a positive value equal to the battery voltage at the drain end to zero voltage at the source end. Since the cross-sectional area of the channel between the gate region is smaller than at either end of the channel (because of the depletion layer), the resistance of this area is relatively large, and most of the voltage drop appears across this part of the channel. The drain end of the channel lying in between the gate region is at a higher potential than the source end of the channel; hence the reverse-bias applied to the p-n junction is greater on this side. The effect on the depletion layer is shown in Fig. 4.3.

When the drain-source voltage is zero (Fig. 4.3*a*), the depletion layer either side of the p-n junction is narrow and has little effect on the channel resistance. Increasing the drain-source voltage above zero will widen the depletion layer and cause it to extend into the channel. This is shown in Fig. 4.3*b* which makes it clear that the layer widens more rapidly at the drain end of the channel than at the source end.

Thus, increasing the drain-source voltage increases the channel resistance and this results in the increase in the drain current being less than proportional to the voltage, i.e. doubling the drain-source voltage does not give a two-fold increase in drain current because the channel resistance has increased also. Further increase in the drain-source voltage

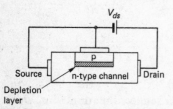

Fig. 4.2 The basic junction fet

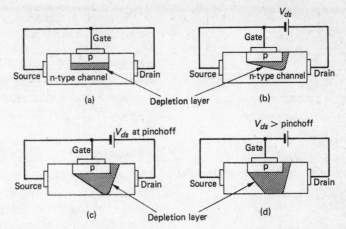

(a)   Depletion layer   (b)

(c)   Depletion layer   (d)

Fig. 4.3 Showing the effect of increasing the drain-source voltage

makes the depletion layer extend further into the channel and eventually the point is reached where the depletion layer extends right across the channel (Fig. 4.3c). The drain-source voltage which produces this effect is known as the PINCH-OFF VOLTAGE. Once pinch-off has developed, further increase in drain-source voltage widens the pinched-off region (Fig. 4.3d). The drain current ceases to increase in proportion to any increase in the drain-source voltage, but is now more or less constant with change in drain-source voltage. The drain current continues to flow because a relatively large electric field is set up across the depleted region of the channel, and this field aids the passage of electrons through the region.

The reverse-bias voltage applied to the gate-channel p-n junction can also be increased by the application of a negative potential, relative to source, to the gate terminal. If the gate-source voltage is made negative, the reverse bias on the gate-channel junction is increased. This increase in bias voltage widens the depletion layer over the width of the gate region and thereby increases the channel resistance. The drain current therefore falls as the gate-source voltage is made more negative until it is approximately equal to the pinch-off voltage and the channel is pinched-off (Fig. 4.4). When this occurs the drain current is zero. Generally, the junction fet is operated with voltages applied to both the drain and the gate terminals, with the drain voltage greater than the pinch-off value. The resistance of the channel up to the pinch-off point is determined by the gate-source voltage, and the drain-source voltage produces an electric field which sweeps electrons across the extended depletion layer. The drain current is then more or less independent of the drain-source voltage and under the control of the gate-source voltage.

A p-channel junction fet operates in a similar manner except that it is necessary to increase the gate-source voltage

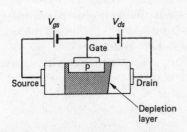

Fig. 4.4 Showing the effect of increasing the gate-source voltage

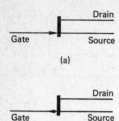

Fig. 4.5 Symbols for (a) an n-channel jfet and (b) a p-channel jfet

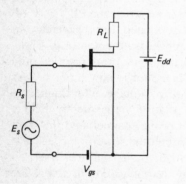

Fig. 4.6 The basic jfet amplifier

in the positive direction to reduce the drain current. Also, of course, the drain is held at a negative potential with respect to the source. The symbols used for n-channel, and p-channel, junction fets are given, respectively, in Figs. 4.5a and b. Both types of junction fet are operated with their gate-channel p-n junction reverse biased; hence they have a very high input impedance.

Application in an Amplifier Circuit

If a junction fet is to operate in an amplifier circuit it must be possible to control the drain current by means of the signal voltage. If the drain current is then passed through a resistance, an output voltage will be developed across the drain resistance that is an amplified version of the input signal voltage. The necessary control of the drain current can be obtained by connecting the signal voltage in the gate-source circuit of the fet (Fig. 4.6).

The signal source, of e.m.f. E_s and impedance R_s, is connected in the gate-source circuit of the fet in series with a bias battery of e.m.f. V_{gs}. The total reverse bias voltage applied to the gate-channel junction is the sum of the signal voltage E_s, and the bias voltage V_{gs}. During the positive half-cycles of the signal waveform, the reverse junction bias is reduced, the depletion layer becomes narrower, and so the drain current increases. Conversely, negative half-cycles of the signal waveform augment the bias voltage and cause the depletion layer to extend further into the channel; the drain current is therefore reduced. In this way, the drain current is caused to vary with the same waveform as the input signal voltage.

The output voltage is developed across the drain load resistor R_L, and can be taken off from between the drain and earth. A voltage gain is achieved because the alternating component of the voltage across R_L is larger than the signal voltage E_s. An increase in the signal voltage in the positive direction produces an increase in the drain current and hence an increase in the voltage developed across R_L. The drain-source voltage V_{ds} is the difference between the drain supply voltage E_{dd} and the voltage across R_L; thus an increase in drain current makes the drain-source voltage fall. This means that a junction fet amplifier operated in the common-source configuration has its input and output signal waveforms in antiphase with one another. It is necessary to ensure that the signal voltage is not large enough to take the gate-source voltage positive by more than about 0.5 V, otherwise the high input-impedance feature of the junction fet will be lost.

The construction of a junction field-effect transistor is shown in Fig. 4.7g.

The various steps involved in the manufacture of an n-channel junction fet are shown by Figs. 4.7a through to g. A heavily-doped p-type silicon substrate marked as p^+ in Fig. 4.7a has a layer of silicon dioxide grown onto its surface (Fig. 4.7b). Next (Fig. 4.7c) a part of the silicon dioxide layer is etched away to create an exposed area of the p-type silicon substrate into which n-type impurities can be diffused. An n-type region is thus produced in the p-type substrate and then another layer of silicon dioxide is grown onto the surface (Fig. 4.7d). The next steps, shown by Fig. 4.7e, are first to etch another gap in the silicon dioxide layer and then to diffuse a p^+ region into the exposed area of the n-type region of the substrate. A third layer of silicon dioxide is then grown over the surface of the device (Fig. 4.7f). Gaps are now etched into the layer into which aluminium contacts to the two ends of the n-type region and the upper p-type region can be deposited (Fig. 4.7g). The terminals connected to the two ends of the n-type region are the source and the drain contacts, while the third terminal acts as the gate.

Parameters

The important parameters of a junction fet are its mutual conductance g_m, its input resistance R_{IN}, and its drain-source resistance r_{ds}. The mutual conductance is defined as the ratio of a change in the drain current to the change in the gate-source voltage producing it, with the drain-source voltage maintained constant, i.e.

$$g_m = \frac{\delta I_d}{\delta V_{gs}} \qquad V_{ds} \text{ constant} \tag{4.1}$$

The drain-source resistance r_{ds} is the ratio of a change in the drain-source voltage to the corresponding change in drain current, with the gate-source voltage held constant, i.e.

$$r_{ds} = \frac{\delta V_{ds}}{\delta I_d} \qquad V_{gs} \text{ constant} \tag{4.2}$$

Typically, g_m has a value lying in the range of 1 to 7 mS, while r_{ds} may be 40 kΩ to 1 MΩ. The input impedance of a junction fet is the high value presented by the reverse-biased gate-channel p-n junction. Typically, an input impedance in excess of $10^8\ \Omega$ may be anticipated.

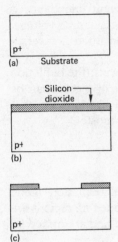

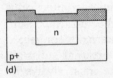

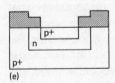

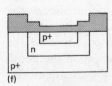

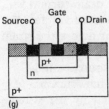

Fig. 4.7 Construction of an n-channel jfet

The Mosfet

The metal-oxide-semiconductor field-effect transistor, generally known as the mosfet, differs from the junction fet in that its gate terminal is insulated from the channel by a layer of silicon dioxide. The layer of silicon dioxide increases the input impedance of the fet to an extremely high value, such as $10^{10} \, \Omega$ or even more. The high value of input impedance is maintained for all values and polarities of gate-source voltage, since the input impedance does not depend upon a reverse-biased p-n junction.

The mosfet is available in two different forms: the depletion type and the enhancement type. Both types of mosfet can be obtained in both n-channel and p-channel versions, so that there are altogether four different kinds of mosfet.

Depletion-type Mosfet

The constructional details of an n-channel depletion mode mosfet are shown in Fig. 4.8. Two heavily doped n^+ regions are diffused into a lightly doped p-type substrate and are joined by a relatively lightly-doped n-type channel.

The gate terminal is an aluminium plate that is insulated from the channel by a layer of silicon dioxide. A connection is also made via another aluminium plate to the substrate itself. In most mosfets the substrate terminal is internally connected to the source terminal but sometimes an external substrate connection is made available. The substrate must always be held at a negative potential relative to the drain to ensure that the channel-substrate p-n junction is held in the reverse-biased condition. This requirement can be satisfied by connecting the substrate to the source. A depletion layer will extend some way into the channel, to a degree that depends upon the magnitude of the drain-source voltage. Because of the voltage dropped across the channel resistance by the drain current, the depletion layer extends further across the part of the channel region nearest to the drain than across the part nearest the source. The resistance of the channel depends upon the depth to which the depletion layer penetrates into the channel. With zero voltage applied to the gate terminal the drain current will, at first, increase with increase in the drain-source voltage, but once the depletion layer has extended right across the drain end of the channel the drain current becomes, more or less, constant with further increase in the drain-source voltage.

The channel resistance, and hence the drain current, of a depletion-type mosfet can also be controlled by the voltage applied to the gate. A positive voltage applied to the gate will attract electrons into the channel from the heavily-doped n^+

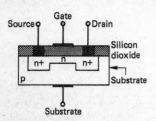

Fig. 4.8 Construction of an n-channel depletion-type mosfet

regions at either end. The number of free electrons available for conduction in the channel is increased and so the channel resistance is reduced. The reduction in channel resistance will, of course, allow a larger drain current to flow when a given voltage is maintained between the drain and source terminals. An increase in the positive gate voltage will increase the drain current which flows when the drain-source voltage is large enough to extend the depletion layer across the drain end of the channel. Conversely, if the gate is held at a negative potential relative to the source, electrons are repelled out of the channel into the n^+ regions. This reduces the number of free electrons which are available for conduction in the channel region and so the channel resistance is increased. The drain current that flows when the depletion layer has closed the channel depends upon the channel resistance.

The drain current of a depletion-type mosfet can therefore be controlled by the voltage applied between its gate and source terminals.

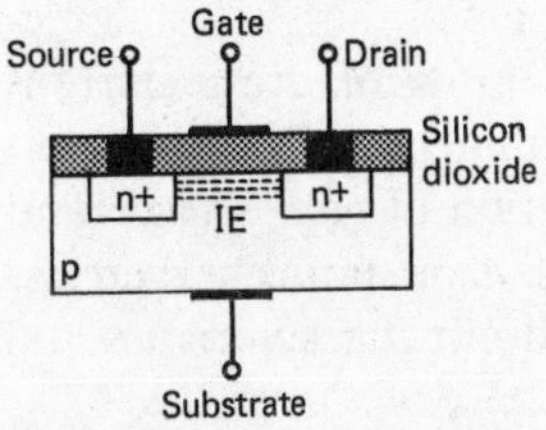

Fig. 4.9 Construction of an n-channel enhancement-type mosfet. IE = induced electrons forming a virtual channel when gate voltage is positive

Enhancement-type Mosfet

Fig. 4.9 shows the construction of an enhancement-type mosfet. The gate terminal is insulated from the channel by a layer of silicon dioxide, and the substrate and source terminals are generally connected together to maintain the channel-substrate p-n junction in the reverse-biased condition. It can be seen that a channel does not exist between the n^+ source and drain regions; hence the drain current that flows when the gate-source voltage is zero is very small. If, however, a voltage is applied between the gate and source terminals, which makes the gate positive with respect to the source, a *virtual channel* will be formed. The positive gate voltage attracts electrons into the region beneath the gate to produce an n-type channel (as shown in the figure) in which a drain current is able to flow. The positive voltage that must be applied to the gate to produce the virtual channel is called the *threshold voltage* and is typically about 2 V. Once the virtual channel has been formed, the drain current which flows depends upon the magnitude of both the gate-source and drain-source voltages. An increase in the gate-source voltage above the threshold value will attract more electrons into the channel region and will therefore reduce the resistance of the channel. The drain current produces a voltage drop along the channel and as with the other types of fet, pinch-off will occur for a particular value of drain voltage. For a particular value of gate-source voltage the drain current will increase with in-

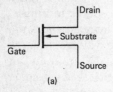

(a)

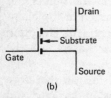

(b)

Fig. 4.10 Symbols for (a) an n-channel depletion-type mosfet and (b) an n-channel enhancement-type mosfet

crease in drain-source voltage up to onset of pinch-off and thereafter will remain more or less constant.

The drain current of a mosfet can hence be controlled by the voltage applied between its gate and source terminals and, if the drain current is passed through a suitable resistance, a voltage gain can be provided. The basic arrangement of a mosfet amplifier is similar to the junction fet circuit given in Fig. 4.6 and it operates in a similar manner.

The important parameters of a mosfet are the same as those of a junction fet: namely, its mutual conductance g_m, its drain-source resistance r_{ds}, and its input resistance R_{IN}. Typically, g_m is in the range 1–10 mS, r_{ds} is some 5–50 kΩ, and R_{IN} is $10^{10}\,\Omega$ or more. It should be noted that whereas the values of mutual conductance are approximately the same as those of a junction fet, the drain-source resistance values are lower but the input resistance is higher.

Figs. 4.10a and b show the symbols for n-channel depletion type and enhancement-type mosfets. The symbols for the p-channel versions differ only in that the direction of the arrow-head is reversed.

Static Characteristics

The static characteristics of a fet are plots of drain current against voltage and are used to determine the drain current which flows when a particular combination of gate-source and drain-source voltage are applied. Two sets of static characteristics are generally drawn: these are the drain characteristics and the mutual characteristics.

Drain Characteristics

The drain characteristics of a fet are plots of drain current against drain-source voltage for constant values of gate-source voltage. The characteristics can be determined with the aid of a circuit such as that shown in Fig. 4.11 for the measurement of the characteristics of an n-channel junction fet.

The data required to plot the drain characteristics consists of the values of the drain current which flows as the drain-source

Fig. 4.11 Circuit for the determination of the static characteristics of an n-channel jfet

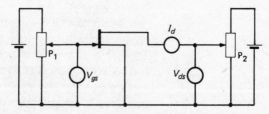

voltage is increased in a number of discrete steps starting from zero, the gate-source voltage being held constant at a convenient value. The method generally used to obtain the data is as follows: the gate-source voltage is set to a convenient value by means of the potential divider P_1 and then the drain-source voltage is increased, starting from zero, in a number of discrete steps. At each step the drain current flowing is noted. The gate-source voltage is then set to another convenient value and the procedure is repeated. In this way sufficient data can be obtained to plot a family of curves of drain current to a base of the drain-source voltage. This family of curves is known as the drain characteristics of the fet. The drain characteristics of the other types of fet are obtained in a similar manner. Fig. 4.12 shows typical drain characteristics for the six types of fet.

It should be noted that each curve has a region of small values of V_{ds} in which I_d is proportional to V_{ds}. In these

Fig. 4.12 The drain characteristic of
(a) an n-channel jfet,
(b) a p-channel jfet,
(c) an n-channel depletion type mosfet,
(d) a p-channel depletion type mosfet,
(e) an n-channel enhancement-type mosfet,
(f) a p-channel enhancement-type mosfet

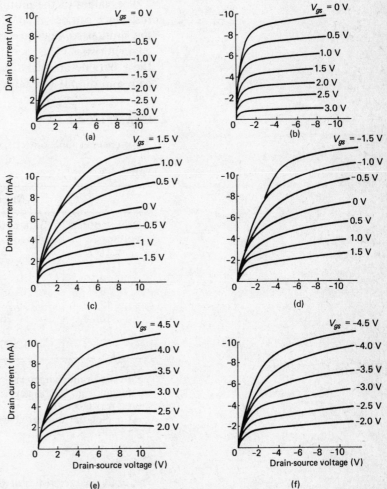

regions the devices can be operated as a voltage-dependent resistance, i.e. as a resistance whose value V_{ds}/I_d depends upon the value of V_{ds}. For all devices the drain current which flows when the gate-source voltage is zero is labelled as I_{dss}.

Mutual Characteristics

The mutual or transfer characteristics of a fet are plots of drain current against gate-source voltage for various constant values of drain-source voltage. The mutual characteristics of an n-channel junction fet can be determined using the arrangement given in Fig. 4.11 and the following procedure. The drain-source voltage is maintained at a constant value as the gate-source voltage is increased in a number of discrete steps. At each step the value of the drain current flowing is noted. The procedure should then be repeated for a number of other drain-source voltages.

The values of the mutual conductance and the drain-source resistance can be obtained from the drain characteristics, while the mutual conductance can be determined from the mutual characteristics. The method employed to obtain the values of these parameters is the same as that to determine the current gain and output resistance of a bipolar transistor.

EXAMPLE 4.1

An n-channel junction fet has the data given in Table 4.1.

Table 4.1

Drain-source voltage V_{ds} (V)	Drain current (mA)			
	Gate-source voltage $V_{gs} = 0$ V	$= -1$ V	$= -2$ V	$= -3$ V
0	0	0	0	0
4	0	5.0	2.4	0.30
8	10.1	5.9	2.7	0.35
12	10.2	6.2	2.9	0.40
16	10.25	6.3	3.0	0.45
20	10.3	6.35	3.05	0.50
24	10.35	6.4	3.1	0.55

Plot the drain characteristics and use them to determine the mutual conductance g_m of the device at $V_{ds} = 12$ V. Calculate also the drain-source resistance for $V_{gs} = -2$ V.

Also plot the mutual characteristics and from them obtain g_m at $V_{ds} = 12$ V.

Solution

The drain characteristics of the fet are shown in Fig. 4.13.

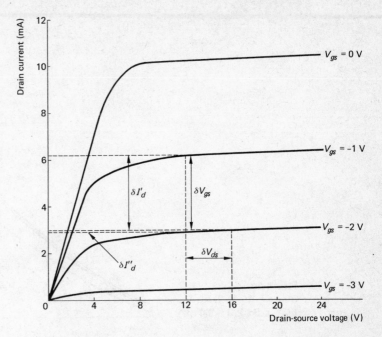

Fig. 4.13
$\delta I'_d = 6.2 - 2.9 = 3.3 \text{ mA}$
$\delta V_{gs} = -1 - (-2) = 1 \text{ V}$
$\delta I''_d = 3.0 - 2.9 = 0.1 \text{ mA}$
$\delta V_{ds} = 8 - 2 = 6 \text{ V}$

The mutual conductance g_m of the fet is given by the expression $g_m = \delta I_d / \delta V_{gs}$, with V_{ds} constant at 12 V. It can be seen from the characteristics that a change in V_{gs} from -2 V to -1 V produces a change in I_d from 2.9 to 6.2 mA. Therefore

$$g_m = \frac{6.2 - 2.9}{2 - 1} \times 10^{-3} = 3.3 \text{ mS} \qquad (Ans.)$$

Also from the characteristics it can be seen that a change in V_{ds} from 12 to 16 V, with V_{gs} constant at -2 V, produces a change in I_d from 2.9 to 3.0 mA. Therefore

$$r_{ds} = \frac{16 - 12}{(3.0 - 2.9) \times 10^{-3}} = 40\ 000\ \Omega \qquad (Ans.)$$

The mutual characteristics of the junction fet are shown plotted in Fig. 4.14. The mutual conductance g_m of the device is given by the slope of the curve; thus for V_{ds} constant at 12 V,

$$g_m = \frac{(6.2 - 2.9) \times 10^{-3}}{1} = 3.3 \text{ mS} \qquad (Ans.)$$

Temperature Effects

The velocity with which majority charge carriers travel through the channel is dependent upon both the drain-source voltage and the temperature of the fet. An increase in the temperature reduces the carrier velocity and this appears in the form of a reduction in the drain current which flows for given gate-source and drain-source voltages.

A further factor that may also affect the variation of drain current with change in temperature is the barrier potential

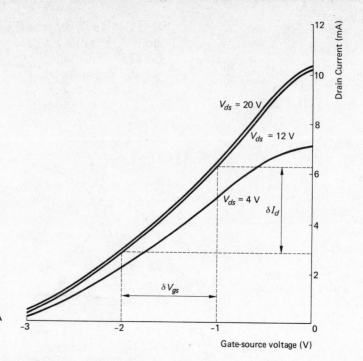

Fig. 4.14
$\delta V_{gs} = -1 - (-2) = 1\,V$
$\delta I_d = 6.3 - 2.9 = 3.4\,mA$

across the gate-channel p-n junction. An increase in temperature will cause the barrier potential to fall and this, in turn, will reduce the width of the depletion layer for a given gate-source voltage. The channel resistance will fall and the drain current will increase. The two effects tend to vary the drain current in opposite directions and as a result the overall variation can be quite small. Indeed, it is possible to choose a particular gate-source voltage and obtain zero temperature coefficient. In general, the overall result is that the drain current decreases with increase in temperature. This is the opposite of the collector current variation experienced by the bipolar transistor.

Handling the Mosfet

The gate terminal of a mosfet is insulated from the channel by a very thin ($\simeq 100\,nm$) layer of silicon dioxide, which effectively forms the dielectric of a capacitance. Any electric charge which accumulates on the gate terminal may easily produce a voltage across the dielectric that is of sufficient magnitude to break down the dielectric. Once this happens the gate is no longer insulated from the channel and the mosfet has been destroyed. The charge necessary to damage a mosfet need not be large since the capacitance between the gate and channel is very small and $V = Q/C$. This means that a dangerously high voltage can easily be produced by merely touching the gate leads with a finger or a tool. To prevent damage to mosfets in store or about to be fitted into a circuit it is usual for them to be kept with their gate and source leads short-circuited together. The protective short-circuit can be provided by twisting the leads together, by means of a springy wire clip around the leads, or by inserting the leads into a conductive jelly or grease. The short-circuit must be retained in place while the device is fitted into a circuit, particularly during the soldering process.

Some mosfets are manufactured with a zener diode internally connected between gate and substrate. Normally, the voltage across the diode is too low for it to operate and it has little effect on the operation of the device. If a large voltage should be developed at the gate by a static electric charge, the zener diode will break down before the voltage has risen to a value sufficiently great to cause damage.

The Relative Merits of Bipolar Transistors and Fets

The input impedance of a bipolar transistor depends upon the d.c. collector current it conducts under quiescent conditions, and for the majority of applications it is somewhere in the region of 1000–3000 ohms. If the transistor is biased so that its quiescent collector current is only a few microamperes, an input impedance of a megohm or more can be achieved. The input impedance of a junction fet is very high, with a mosfet having an impedance which is several orders higher still. The mutual conductance of a bipolar transistor is 38 mS per mA of alternating collector current and is therefore considerably higher than the mutual conductance of a fet; this means that the bipolar transistor is capable of providing the larger voltage gain. The collector current of a bipolar transistor increases with increase in temperature and thermal runaway is a possibility unless suitable d.c. stabilization circuitry and/or heat sinks are used. The drain current of a fet decreases with increase in temperature and there is no risk of thermal instability.

When a bipolar transistor is used as a radio-frequency amplifier in a superheterodyne radio receiver, third-order (cubic) terms in its mutual characteristic will lead to an undesirable effect, known as CROSS-MODULATION, whenever a large-amplitude unwanted signal is applied. Cross-modulation is an effect in which the amplitude modulation on the unwanted carrier signal is transferred on to the wanted carrier signal to produce interference. The mutual characteristic of a fet contains these third-order terms at a very much smaller amplitude, if at all, and cross-modulation is not a problem.

When a fet is used as a switch its ON resistance is larger than the ON resistance obtainable from a transistor but the switching operation can be carried out in either direction, i.e. the drain and source terminals are interchangeable. On the other hand, the switching speed of the fet is slower than that of the bipolar transistor. This is because the ON resistance of a fet is larger than that of a bipolar transistor and so a fet is unable to charge or discharge the stray capacitances and the input capacitance of the next stage as quickly.

Exercises

4.1. The drain characteristics of a fet are given in Table A. Plot the characteristics and determine the drain-source resistance from the characteristic for $V_{gs} = 0.5$ V.

Use the curves to find the mutual conductance for $V_{ds} = 20$ V. What type of fet is this?

Table A

	Drain current I_d (mA)				
Drain-source voltage V_{ds} (V)	Gate-source voltage $V_{gs} = 1$ V	0.5 V	0 V	−0.5 V	−1 V
10	4.00	3.19	2.38	1.57	0.76
20	4.02	3.21	2.40	1.59	0.78
30	4.04	3.23	2.43	1.61	0.80

4.2. Draw sketches to show the various steps in the manufacture of an enhancement type mosfet.

4.3. The data given in Table B refer to a fet. Use the data to draw the drain characteristics, and then use the curves to determine: (i) the drain-source resistance for $V_{gs} = -3$ V, and (ii) the mutual conductance for $V_{ds} = -6$ V.

State the type of fet to which the data refer.

Table B

	Drain current (mA)		
Drain-source voltage $V_{ds} = $ (V)	Gate-source voltage $V_{gs} = -4$ V	−3 V	−2 V
−1	−6.3	−4.6	−3.1
−3	−6.6	−5.1	−3.5
−5	−7.2	−5.6	−3.9
−7	−7.8	−6.1	−4.3
−9	−8.4	−6.6	−4.7

4.4. An n-channel junction fet has the data given in Table C. Plot the drain and mutual characteristics and use them to determine the mutual conductance of the device. Find also the drain-source resistance.

Table C

Drain Current (mA)				
Drain-source voltage $V_{ds} = $ (V)	Gate-source voltage $V_{gs} = 0$ V	−0.5 V	−1.0 V	−1.5 V
10	2.25	1.35	0.7	0.3
20	2.29	1.38	0.73	0.33
30	2.32	1.41	0.75	0.35

4.5. Explain, with the aid of a circuit diagram, how you would measure the drain and mutual characteristics of an n-channel mosfet. Draw a typical set of drain characteristics and say how the characteristics of a p-channel mosfet would differ from those shown.

4.6. Describe, with the aid of sketches, the principles of operation of a p-channel junction field-effect transistor. Sketch a typical family of drain characteristics and account for their shape.

4.7. The drain characteristics of a field-effect transistor are as given in Fig. 4.15. Use the characteristics to obtain the values of drain current corresponding to various values of gate-source voltage when V_{ds} is (i) −3 V and (ii) −6 V and thence plot the mutual characteristics of the transistor for these drain-source voltages.

4.8. The relationship between the drain current and the drain-source voltage of a field-effect transistor up to pinch-off is given by Table D.

Table D

V_{ds} (V)	0.5	1.0	1.5	2.0	2.5	3.0	3.5	4.0
I_d (mA)	0	5	7	8	11	12.5	14.7	16

Plot the characteristic over this range and thence determine the a.c. resistance of the transistor when (i) $V_{ds} = 1.0$ V, (ii) $V_{ds} = 2.0$ V, and (iii) $V_{ds} = 3.0$ V. Suggest a possible use of the device when operated on this part of its characteristics.

4.9. Explain the meaning of the following terms used in conjunction with fets and draw characteristic curves and constructional sketches as necessary to illustrate your answers:
(i) n-type channel, (ii) junction gate, (iii) pinch-off region, (iv) pinch-off voltage, (v) metal oxide semiconductor gate, (vi) threshold voltage (C&G)

4.10. Describe the principle of operation, and give the symbol, of an enhancement-type mosfet. Illustrate your answer with typical drain characteristics.

Short Exercises

4.11. A depletion-type mosfet can be used in either the enhancement or the depletion mode. Explain why this is not also true for a junction fet or for an enhancement-type mosfet.

4.12. Compare the properties of a field-effect transistor with those of a bipolar transistor.

4.13. State the precautions that must be taken when using a mosfet.

4.14. Draw the symbol for each type of field-effect transistor. How

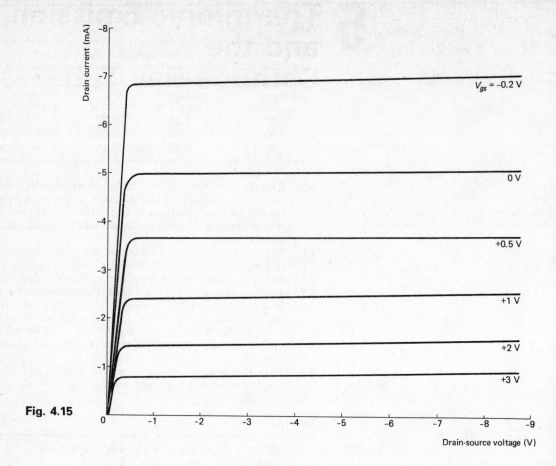

Fig. 4.15

do the symbols show (i) the use of p or n channel, (ii) the use of an insulated gate, (iii) whether or not a conducting path exists between drain and source when the gate-source voltage is zero?

4.15. Compare the relative merits of field-effect and bipolar transistors as electronic switches.

4.16. Why may an increase in temperature lead to thermal runaway in a bipolar transistor but not in a field-effect transistor?

4.17. (a) In which direction do majority charge carriers flow in an n-channel jfet?

(b) A fet is operated as an amplifier with a positive drain supply voltage and its gate held positive relative to its source. What kind of fet is it?

(c) What is meant by pinch-off?

4.18. If the gate of an n-channel jfet is taken positive with respect to the source by more than about 0.5 V the high input impedance feature of the device is lost. Explain the reason for this.

4.19. Sketch the construction of a p-channel jfet.

4.20. What is thermal runaway? Why is it a problem with bipolar transistors but not with fets?

4.21. (a) How are some mosfets internally protected against damage caused by touching the gate terminal?

(b) What is the effect of this protection on the input impedance of the device?

5 Thermionic Emission and the Cathode Ray Tube

A large number of different types of valve have been used in the past but nowadays the vast majority of circuit functions are carried out using transistors and/or integrated circuits. The main application remaining to valve technology lies in very high power circuitry such as high-power radio transmitters. These use either *triodes* or *tetrodes* in their final stage, and sometimes in their penultimate stage also. In older equipments another type of valve, known as the *pentode*, was widely employed.

In all these types one of the electrodes is the cathode and another is the anode, and the basic principle of operation is that when the cathode is heated to a suitable temperature it emits large numbers of electrons. A proportion of these are collected by the anode and constitute the anode current. In their passage from cathode to anode, the emitted electrons may pass through one, two or three grids—depending on the type of valve—and potentials applied to these grids are able to control the flow of electrons and hence the anode current.

Thermionic Emission

At room temperatures the electrons in a metal are able to wander at random in the atomic structure of the metal, and some of the electrons near to the surface of the metal will pass out into the surrounding air. At room temperatures ordinary metals do not lose their electrons in large numbers and so a force must exist that prevents electrons from permanently leaving the metal's surface. Immediately an electron leaves the metal the metal has lost a negative charge (the electronic charge) and this is equivalent to its gaining a positive charge. The positive charge exerts a force on the emitted electron that pulls the electron back towards the metal, and for an electron to be able to escape it must possess sufficient kinetic energy to overcome this force. At room temperatures very few electrons have sufficient energy and the number of electrons able to escape from the metal is very small. In order to considerably increase the number of electrons escaping from the metal it is necessary to give the electrons additional energy, and this can best be done by heating the metal. As the temperature of the metal is increased more electrons attain sufficient energy to enable them to overcome the retarding force and are able to escape from the metal.

For most metals the temperature necessary for adequate electron emission is too high if the characteristics of the metal are not to be changed. In practice, the choice of cathode material is limited to tungsten, thoriated tungsten and nickel coated with a mixture of barium oxide and strontium oxide to which some calcium oxide may also be added.

Once an electron has escaped from the cathode it is subjected to a retarding force set up by the negatively charged electrons that have already escaped. An electron is slowed down by this force and may well be returned to the cathode. An emitted electron can also be slowed down by coming into collision with a gas molecule but to minimize this effect the cathode is placed within an evacuated glass envelope.

Thus electrons are continually emitted from the surface of a heated cathode and are then subjected to forces which tend to return them to the cathode. Most of the emitted electrons travel only a short distance from the cathode before they are returned to it; this distance is proportional to the velocity with which an electron is emitted. The cathode is surrounded, therefore, by a cloud of electrons, some of which are moving away from the cathode and some moving towards it. Only those electrons having sufficient energy to overcome the restraining forces can escape from the vicinity of the cathode. The cloud of electrons around the cathode is known as the SPACE CHARGE. Obviously the space charge is negative.

If another electrode, the anode, is introduced into the evacuated space and is maintained at a positive potential with respect to the cathode, it will exert an attractive force on the emitted electrons. The emitted electrons are then subjected to a retarding force produced by the electric fields set up by the positively charged cathode and the negatively charged space charge, and an attractive force produced by the electric field set up by the anode. Near to the anode the accelerating electric field is larger than the retarding electric field, while near to the cathode the retarding electric field predominates. At some point in between the cathode and the anode the two electric fields are equal and cancel out, and an electron reaching this point is neither accelerated nor retarded. Any emitted electrons that have sufficient energy to reach this point come under the influence of the anode field and are rapidly accelerated to the anode and there contribute to the anode current.

An increase in the positive potential of the anode moves the point of zero electric field nearer to the cathode and allows electrons of lower energy to reach the anode. The anode current thus increases with increase in anode voltage. If the anode voltage is increased to the point where the position of zero electric field occurs at the cathode, all the emitted electrons will reach the anode and the space charge is then non-existent. The anode current is then at its maximum value, since there are no more emitted electrons available for collection by the anode. A further increase in anode voltage does not produce a corresponding increase in anode current; an increased anode current can only be obtained by increasing the cathode temperature and thus increasing the number of emitted electrons.

A valve with just a cathode and an anode is known as a diode. A triode valve has a third electrode, known as the control grid, inserted in between the cathode and the anode. If this electrode is held at a negative potential with respect to the cathode, it will exert considerable control over the anode current. This means that, if a varying voltage is applied to the grid, it will produce a larger varying voltage at the anode of the valve, i.e. amplification is provided.

The tetrode valve has another grid inserted between the control grid and the anode while the pentode valve has three grids in all.

The CATHODE RAY TUBE (c.r.t.) is a thermionic device that is able to display visually the instantaneous values of electrical signals. A beam of high-velocity electrons is directed on to a screen coated with a fluorescent substance and which therefore shows a visible spot of light. By deflecting the electron beam, the visible spot can be moved about the screen and an electrical waveform described. The cathode ray tube is widely employed in electronic and telecommunication engineering; the most evident example of its use is the television receiver in the home, and the other example of importance to this book is the cathode ray oscilloscope (c.r.o.) which is an electrical measurement instrument.

Essentially, a cathode ray tube consists of the source of a high-velocity electron beam, a means of focussing the electron beam, and a means of deflecting the beam about the face of a fluorescent screen. Electric and magnetic fields can be utilized for the focussing and deflection of the beam, and each possess relative advantages and disadvantages which lead to their respective adoption for tubes designed for different applications.

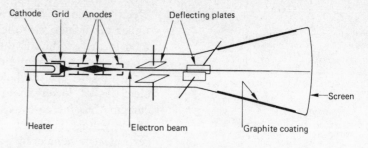

Fig. 5.1 The cathode ray tube with electric focussing and electric deflection

Electric Focussing and Electric Deflection

The basic construction of a c.r.t. which uses electric fields for both focussing and deflection of the electron beam is shown in Fig. 5.1. The tube consists of an evacuated glass envelope inside which are positioned an electron gun, two pairs of deflecting electrodes, and a screen whose inside surface is coated with a fluorescent material. The component parts of the electron gun are an indirectly heated cathode, a cylindrical modulator or grid that surrounds the cathode, and three (sometimes two) anodes. The grid is maintained at a negative potential, and each anode at a positive potential, relative to the cathode. The electrons emitted from the cathode are subjected to a repulsive force exerted by the negative grid and to an attractive force applied by the anodes.

The positive potentials applied to the anodes are much larger than the negative potentials applied to the grid (typically +1000 V as opposed to −20 V), but since the grid is much nearer to the cathode than are the anodes, its influence is sufficiently great for it to control the number of electrons that pass through the small hole in the grid. These electrons form the electron beam. The electrons leaving the grid are accelerated to a high velocity by the positive anode potentials and travel along the tube until they strike against the screen. When the electrons impact upon the screen, the fluorescent material with which the screen is coated glows visibly, and a spot of light can then be seen on the face of the screen. The brightness of this spot of light depends upon the number of electrons reaching the screen and their velocity, and these factors are under the control of the grid potential. If, therefore, the grid potential is provided by a potential divider connected across a suitable voltage supply, a BRILLIANCE CONTROL can be provided (Fig. 5.2).

Since each of the emitted electrons carries a negative charge they have a mutually repulsive effect upon one another, causing the beam to diverge as it travels along the tube (Fig. 5.3). Such a beam would produce a faint and blurred area of light on the face of the screen. If electrical waveforms are to be observed, a small and sharply focussed spot of light should be

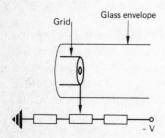

Fig. 5.2 Brilliance control

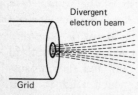

Fig. 5.3 Electron beam diverging as it leaves the grid

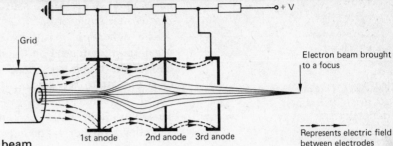

Fig. 5.4 Focussing the electron beam

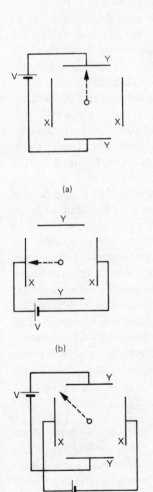

Fig. 5.5 Principle of electric deflection of the electron beam

visible on the screen and so some means of focussing the electron beam is necessary. The electric field set up between the grid and the first anode exerts forces on the electrons that bring them to a focus at a point just in front of the anode. There the beam diverges and is brought to a second focus at the end of the tube by the electric fields set up in the region in between the three anodes. The focal point can be made to coincide with the position of the screen by suitable adjustment of the potential applied to the middle anode by means of the FOCUS CONTROL. Fig. 5.4 shows how the electron beam is focussed.

Sometimes the required potential differences between the cathode and the anodes is obtained with the cathode held at a highly negative potential and the anodes at zero potential or a small positive potential.

Situated between the focussing system and the screen are two pairs of beam-deflecting plates. The X-plates are mounted vertically and provide a horizontal electric field to deflect the electron beam in the horizontal plane. The horizontally mounted plates provide deflection in the vertical plane. Fig. 5.5 illustrates the principle of electric deflection of the electron beam.

When a voltage is applied across the X-plates (Fig. 5.5b), the beam will be deflected in the horizontal plane towards the plate that is held at the positive potential. The distance through which the beam is moved is proportional to the magnitude of the plate voltage. The DEFLECTION SENSITIVITY is the voltage that must be applied across the plates to produce 1 cm deflection of the visible spot on the screen. Similarly, if a voltage is applied across the Y-plates the spot will be vertically deflected (Fig. 5.5a). If voltages are simultaneously applied across both the X-plates and the Y-plates, as in Fig. 5.5c, the electron beam is subjected to deflecting forces exerted in two different directions at the same time and is therefore deflected in the resultant direction. Thus, if the two voltages are of equal amplitude, and the X- and Y-plates have equal deflection sensitivities, the spot will be deflected through an angle of 45°.

The deflection sensitivity is inversely proportional to the accelerating potential on the third anode but, since the brightness of the visible spot depends upon the velocity with which electrons strike the screen, the anode voltage cannot be reduced very much in an attempt to increase the deflection sensitivity. Some cathode ray tubes overcome this difficulty by including a further anode between the deflecting plates and the screen to provide *post-deflection acceleration*. This extra anode consists of a graphite coating on the inside wall of the tube and it is held at the same potential as the third anode.

The impact of high-velocity electrons on to the screen causes the screen to glow visibly, and the colour of the display depends upon the fluorescent material used to coat the screen. Green is generally considered to be the best colour for the tube in an oscilloscope and to obtain it zinc orthosilicate is used. The incident electrons have sufficient kinetic energy to cause the screen to emit secondary electrons.

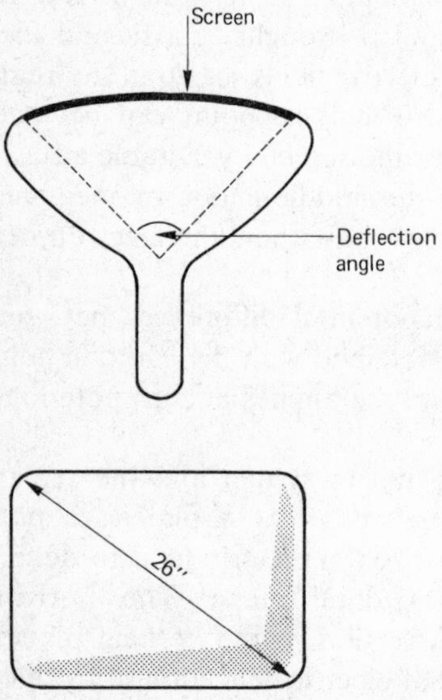

These electrons must be collected and returned to the cathode, otherwise the number of electrons present on the screen will continually increase and eventually cause the screen to become negatively charged. Usually the secondary electrons are collected by a graphite coating on the inside of the neck of the tube which is connected indirectly to the cathode.

A c.r.t. with electric focussing and electric deflection is sensitive and can operate at high frequencies but the maximum possible deflection angle (see Fig. 5.6a) is not large. Because of these factors this type of tube is used in cathode ray oscilloscopes where sensitivity and high maximum frequency are of greater importance than the possible deflection angle. The screen of an oscilloscope tube is therefore always rather small, typically 7 in. Conversely, in the television receiver the size of the screen (Fig. 5.6b), is always larger, typically 22 in. or 26 in., and for such sizes a large deflection angle is essential. Tubes for use in television receivers do not, therefore, employ electric deflection of the electron beam.

Fig. 5.6 Showing (a) the deflection angle of a cathode ray tube and (b) how the size of a screen is measured

Magnetic Focussing and Magnetic Deflection

The basic constructional details of a cathode ray tube which uses both magnetic focussing and magnetic deflection of the electron beam is shown in Fig. 5.7. Electrons are emitted from the heated cathode and pass through the hole in the cylindrical grid to be accelerated to a high velocity by the positive potentials at the anodes.

The electric field set up between the grid and the first anode focusses the electron beam at a point near the anode, beyond

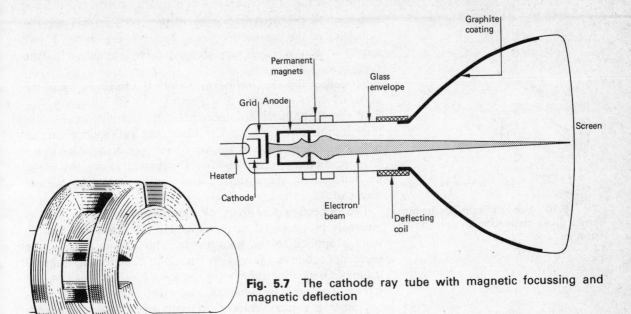

Fig. 5.7 The cathode ray tube with magnetic focussing and magnetic deflection

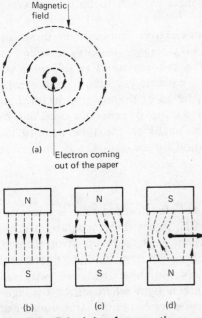

Fig. 5.8 Magnetic deflection coils

Fig. 5.9 Principle of magnetic deflection

which the beam starts to diverge. The permanent magnets mounted around the outside of the tube produce a magnetic field through which the divergent electron beam passes. Magnetic forces are exerted upon the beam and have the effect of bringing it to a focus at the screen. Fine adjustment of the focal point is possible by moving the physical position of the magnets and thereby altering the magnetic field set up inside the tube. When the magnets are moved closer together, the magnetic field they produce is increased and the beam is then focussed at a point nearer to the magnets. Similarly, increasing the magnet spacing will move the focal point of the beam further away from the position of the magnets. The electron beam then passes into the magnetic field set up by two pairs of deflecting coils wound on an insulating former positioned over the neck of the tube (see Fig. 5.8).

The way in which a magnetic field is able to deflect an electron beam is illustrated by Fig. 5.9. A moving electron constitutes an electric current and so a magnetic field will exist around it, the direction of which can be determined using Maxwell's Corkscrew Rule (Fig. 5.9a). Fig. 5.9b shows the direction of the magnetic field that exists between two current-carrying coils; one with North magnetic polarity, the other with South. If an electron moves into the magnetic field of Fig. 5.9b the magnetic fields will interact and, since magnetic lines of force always try to reduce their length, a force is exerted upon the electron moving it to the left (Fig. 5.9c). If the

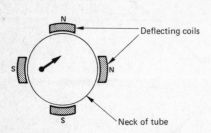

Fig. 5.10 Use of four deflecting coils gives deflection in any direction

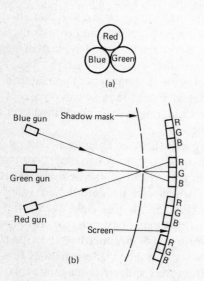

Fig. 5.11 Principle of the shadow mask colour tube

direction of the current flowing in the coils is reversed, the polarities of the magnetic poles of the coils are changed over and the electron is thus subjected to a force that moves it to the right (Fig. 5.9d).

To deflect the electron beam in both the horizontal and the vertical planes, two pairs of deflecting coils are necessary. Fig. 5.10 shows the basic arrangement. If the currents in the coils flow in the directions and produce the magnetic polarities indicated, the electron beam will be subjected to forces trying to move it both upwards and to the right, and it will, therefore, be deflected in the resultant direction as shown by the arrowhead.

The deflection sensitivity of a magnetic deflection tube is inversely proportional to the square root of the accelerating voltage applied to the final anode. This means that for the same deflection sensitivity as an electric deflection tube, a magnetic deflection tube will produce a brighter trace on the screen. To obtain an adequate deflecting magnetic field without passing a large current through the windings, the coils must have quite a large inductance. The reactance of these windings will increase in frequency, reducing the current flowing in them and therefore also the strength of the deflecting magnetic field produced. The deflection sensitivity, therefore, falls at higher frequencies and so the use of a magnetic deflection system is restricted to cathode ray tubes that will not be called upon to display high-frequency waveforms. The main application for magnetic deflection used to be the television receiver.

Magnetic focussing of the electron beam has the disadvantage of requiring rather a bulky magnet assembly mounted on the outside of the tube. The modern trend in television receiver design is to employ a cathode ray tube having electric focussing and magnetic deflection of the electron beam. Fine focus adjustment for such a tube is generally achieved by means of small magnets, mounted on the tube near to the deflecting coils, whose position can be varied.

Colour Television Tubes

The most common type of colour television tube is known as the SHADOW-MASK TUBE. It has three separate electron guns, each of which directs an electron beam at the screen. The screen is coated with three different kinds of fluorescent material, one of which glows with blue light, one with red light, and the third with green light when struck by high-velocity electrons. The screen is divided into a large number of small areas each of which contains a fluorescent material of each kind as shown in Fig. 5.11a.

The electron beam from a particular gun must only be able to strike screen elements of *one* colour. To achieve this, a perforated steel sheet, known as the shadow-mask, is mounted between the electron guns and the screens, and is carefully positioned, as shown in Fig. 5.11*b*. As the electron beams scan the screen, light of the three different colours is emitted and the eye sees the colour which corresponds to their addition.

Timebases

Alternating voltage waveforms are functions of time and to display such a waveform correctly on an oscilloscope cathode ray tube, the visible spot must be moved with a constant velocity across the face of the screen. When the spot reaches the right-hand side of the screen, it must be returned rapidly to the left-hand side to begin another journey across the screen. To satisfy this requirement, a voltage that rises linearly with time to some maximum value and then falls rapidly to zero must be applied across the X-plates. Such a voltage waveform, known as a SAWTOOTH, is shown in Fig. 5.12 and is produced by a *timebase* circuit. It is desirable that the rapid FLYBACK of the electron beam should not produce a visible trace on the screen; this is done by applying blanking pulses to the grid of the tube during the flyback period. These negative pulses bias the tube beyond cut-off so that no electrons are emitted from the cathode.

If an alternating voltage is applied across the Y-plates, it causes the electron beam to be simultaneously deflected in the vertical plane, and if the periodic times of the timebase and signal voltages are equal, or one is an exact low multiple of the other, the visible spot will describe the correct waveform on the screen. Suppose for example, as in Fig. 5.13, that the signal voltage has a sinusoidal waveform and the timebase waveform has been adjusted to have the same periodic time as the signal. The position of the visible spot on the screen is determined by the resultant deflecting force exerted on the electron beam. Hence, at time instant 1, both voltages are zero

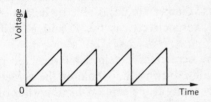

Fig. 5.12 Oscilloscope timebase waveform

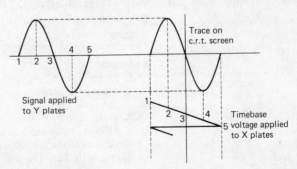

Fig. 5.13 Derivation of trace on c.r.t. screen

and the spot is not deflected. At instant 2 the signal voltage is at its maximum positive value and the timebase voltage has increased in a linear manner to a quarter of its final value; the instantaneous position of the spot is found by projecting from the two waveforms, shown by dotted lines, and noting the point of intersection. In similar fashion the spot position at instants 3, 4 and 5 can be determined. At instant 5 the timebase voltage has reached its maximum value and is then reduced abruptly to zero ready to start another ramp. Because the afterglow of the fluorescent material is short, only a small area of the screen is actually glowing at any instance but, because of the persistence of vision of the human eye, a continuous trace is seen.

The position of the displayed waveform on the screen can be adjusted in both the horizontal and the vertical planes by superimposing a d.c. bias voltage on to the signal and/or the timebase voltage. The bias voltages are adjustable by means of variable resistors, known, respectively, as the X and Y *shift controls*.

The visible spot on the screen of a television receiver is required to trace out a path covering the whole of the screen (see Fig. 5.14). The SCAN starts at the top left-hand corner of the screen, point A, and travels along the first line to point B. At the end of this line the scan rapidly flies back to point C before travelling along the next line to point D, and so on until the entire screen has been scanned. It is usual to employ INTERLACED SCANNING in which the odd lines 1, 3, 5, 7, etc. are scanned first and the even lines afterwards. It is therefore necessary for the electron beam to be deflected in both the horizontal and the vertical planes. Horizontal deflection is provided by the *line scan coils* and vertical deflection by the *field scan coils*. Both pairs of coils must have a sawtooth current waveform passed through them, but while the line scan current is at the high frequency of 15 625 Hz the field scan current is at only 50 Hz (Fig. 5.15).

As the screen is scanned, the brightness of the spot is modulated by the picture signal waveform, which is applied to vary the grid/cathode voltage of the tube. Again because of the persistence of vision of the human eye a complete picture is seen.

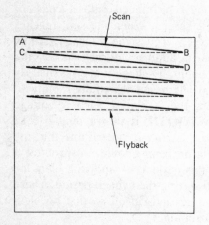

Fig. 5.14 Sequential scanning of a television picture

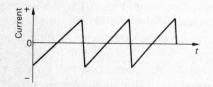

Fig. 5.15 Line and field scan currents

Exercises

5.1. Draw a sketch to show the essential parts of a c.r.t. suitable for use in an oscilloscope. Describe the principle of operation of the tube.

5.2. Draw a sketch showing the construction of a c.r.t. employing electric focussing and magnetic deflection of the electron beam. Give reasons for the use of an electric field for focussing, and a magnetic field for deflecting, the electron beam.

5.3. What is a timebase? Draw the timebase waveform required for the c.r.t. in an oscilloscope and explain (i) why the increase in voltage with time should be linear, (ii) why the flyback period should be short, and (iii) when and why blanking pulses are applied to a c.r.t.

5.4. Describe, with the aid of sketches, how deflection of a beam of electrons can be carried out using (i) a magnetic field and (ii) an electric field. Compare the two methods and state their applications.

5.5. A voltage rising linearly with time from 0 V to 100 V in 1 ms and then falling to zero in negligible time is applied across the X-plates of a cathode ray tube. A sinusoidal voltage of 100 V peak value and frequency 500 Hz is applied across the Y-plates. Assuming equal X and Y plate sensitivities of 25 V/cm, draw the waveform displayed on the screen.

5.6. A particular c.r.t. requires that the potentials of the first, second and third anodes relative to the cathode are, respectively, +600 V, +900 V and +1200 V. The grid bias voltage is −40 V with respect to the cathode. If the cathode is held at a potential of −600 V relative to Earth potential, at what potentials, relative to Earth, are the three anodes and the grid? If the required electrode voltages are to be obtained from the resistive network shown in Fig. 5.16 determine values for the resistors R_1 to R_6. Assume a current of 1 mA in each resistive chain.

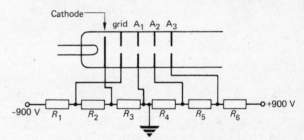

Fig. 5.16

5.7. Explain the principles and practical construction of each of the following systems as applicable to modern c.r.t.s (*a*) an electromagnetic deflection system, (*b*) an electrostatic focussing system, (*c*) an electrostatic deflection system. (C&G)

5.8. (*a*) Give a labelled drawing which clearly illustrates the internal component parts of a monochrome television tube as used in present-day receivers. (*b*) Explain briefly how the electron beam is (i) generated, (ii) modulated, (iii) focussed and (iv) deflected. (C&G)

Short Exercises

5.9. List the relative merits of (i) electric and magnetic focussing and (ii) electric and magnetic deflection of the electron beam in a c.r.t.

5.10. Describe the path taken by the electrons emitted from the cathode of a c.r.t.

5.11. What is meant by the following terms when used in connection with a c.r.t., (i) X shift control, (ii) blanking pulses, (iii) focus control, and (iv) intensity control?

5.12. Why do c.r.t.s designed for use in an oscilloscope use electric deflection of the electron beam while t.v. receiver tubes employ magnetic deflection?

5.13. Draw the diagram of an electron gun and explain its operation.

5.14. The X-plates of a c.r.t. have a sensitivity of 50 V/cm when the final anode voltage is 1000 V. What will be the sensitivity if the anode voltage is increased to 1200 V?

5.15. How is the focussing of the electron beam adjusted in a c.r.t. with (i) electric focussing and deflection, (ii) electric focussing and magnetic deflection, and (iii) magnetic focussing and deflection?

5.16. What is the purpose of (i) the accelerating voltage in a c.r.t. and (ii) the graphite coating along the inside of the glass envelope of a cathode ray tube? Why is more than one anode employed?

5.17. Define the term *deflection sensitivity* as applied to a c.r.t. A particular c.r.t. has an X-plate sensitivity of 25 V/cm. What voltage must be applied to the X-plates to deflect the spot by 3 cm?

6 Small Signal Amplifiers

Field-effect and bipolar transistors are able to amplify a.c. signals because their output current can be controlled by a signal applied to their input terminals. A change in the input signal causes a change in the output current, and if a voltage or power output is required the output current must be passed through a resistive load in the output circuit. Henceforth, the bipolar transistor will be referred to simply as "transistor".

The fet has a very high input impedance and is a voltage-operated device and this means that a fet amplifier can provide a voltage gain only. Voltage gain is defined as the ratio of a change in output voltage to the change in input voltage producing it. The transistor, on the other hand, has a relatively low input impedance and requires an input current for its operation; a transistor amplifier, therefore, is capable of giving a voltage, current, or power gain. Current gain is defined as the ratio of a change in output current to the change in input current producing it, and power gain is defined as

$$\text{Power gain} = \frac{(\text{change in output current})^2 \times \text{load resistance}}{(\text{change in input current})^2 \times \text{input resistance}}$$

Choice of Operating Point

The dynamic mutual characteristic of a fet with a resistive drain load shows how its output current varies with change in its input voltage for particular values of load resistance and supply voltage. Similarly, the dynamic transfer characteristic of a transistor shows how the collector current of the transistor varies with change in the base current for particular values of collector load resistance and collector supply voltage.

A dynamic characteristic can be used to determine, graphically, the waveform of the output current for a particular input signal waveform. Ideally, the two waveforms should be identical, but this requires the dynamic characteristic to be absolutely linear. In practice, some non-linearity always exists and, for minimum signal distortion, care must be taken to restrict operation to the most linear part of the characteristic. For this a suitable operating, or QUIESCENT, point must be selected and the amplitude of the input signal must be limited. The chosen operating point is fixed by the application of a steady bias voltage or current. For maximum signal handling performance the operating point is usually placed in the centre of the linear portion of the dynamic characteristic. Then an alternating signal centred on this operating point produces equal swings of output current above and below the quiescent value, as shown in Figs. 6.1a and b. If maximum undistorted output voltage and/or current swings are not required, the operating point may be chosen to give a small d.c. collector current to minimize demand on the power supply (particularly if a battery is used), or to give the particular value of collector current at which the current gain h_{fe} of the transistor is a maximum.

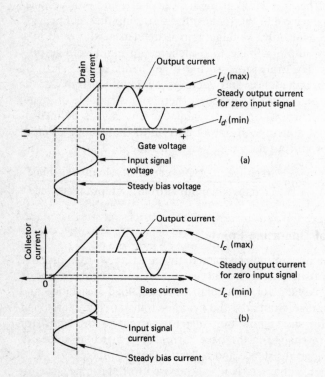

Fig. 6.1 Variations of output current with input signal for (a) a fet and (b) a transistor

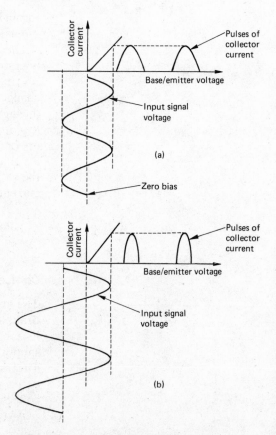

Fig. 6.2 (a) Class B, (b) Class C operation of a transistor

The output current may conveniently be regarded as a direct current having an alternating current superimposed upon it. The direct current is equal to the current that flows when the input signal is zero, i.e. the quiescent current, and the alternating current has a peak-to-peak value of $I_{max} - I_{min}$.

The output current clearly flows at all times during a cycle of the input signal waveform. The active device is said to be operated under CLASS A conditions. The peak value of the signal waveform should, at all times, be less than the bias voltage, or current, otherwise the output waveform will be distorted.

Note that Fig. 6.1b shows the input current as having a sinusoidal waveform; however, the relationship between the input voltage and the input current of a transistor is not a linear one unless the steady bias current is much larger than the signal current. This is because the input resistance of a transistor is not a constant quantity but varies with change in input voltage (refer to Chapter 3).

Class B and Class C Operation

Class A operation of an amplifier offers low signal distortion. The maximum theoretical efficiency with which the d.c. power taken from the power supply is converted into a.c. signal power output is, however, only 50%, and practical efficiencies are lower than this—particularly in the case of thermionic valve amplifiers. To obtain a greater efficiency than this, an amplifier may be operated under either Class B or Class C conditions.

With Class B operation (see Fig. 6.2a) which refers to a transistor, the operating point is set at cut-off. The output current flows only during alternate half-cycles of the signal waveform. It is evident that the output current waveform is highly distorted; Class B bias can therefore only be used with circuits that are able to restore the missing half-cycles of the signal waveform. Such circuits are known as PUSH-PULL amplifiers and TUNED-RADIO FREQUENCY amplifiers. Class B operation has a maximum theoretical efficiency of 78.5%.

Even greater efficiency can be obtained with Class C bias. With Class C bias, shown in Fig. 6.2b, the operating point is set well beyond cut-off. The output current flows in the form of a series of narrow pulses having a duration which is less than half the periodic time of the input signal waveform. Class C bias is used with radio-frequency power amplifiers and with some oscillator circuits. Neither Class B nor Class C bias can be used in conjunction with a resistance loaded audio-frequency amplifier, because of the excessive distortion which would then result.

Fet Bias

Referring to Fig. 4.12 (p. 61) it can be seen that both the junction fet and the depletion-mode mosfet conduct a drain current when their gate-source voltage is zero. This means that one of these fets will operate if a resistor is connected between the gate and earth and the source is taken directly to earth (Fig. 6.3a). Then both the gate and the source are at earth potential and so $V_{gs} = 0$ V. The disadvantage of this simple circuit is that it provides no d.c. stabilization of the drain current and, for a jfet, the input signal voltage must be very small to avoid the gate-channel p-n junction becoming forward biased. Better results are obtained if a resistor R_3 is connected between source and earth (Fig. 6.3b). A bias voltage V_{gs}, equal to the product of the drain current and the source resistance, is developed across R_3. The resistor R_1 is still needed to provide a d.c. path between the gate and earth and this should be some hundreds of kilohms to avoid shunting the signal path.

Improved stabilization of the drain current can be achieved if another resistor R_2 is connected between the drain supply voltage and the gate terminal (Fig. 6.3c). R_1 and R_2 form a potential divider across the drain supply voltage to keep the gate potential constant. The gate-source voltage V_{gs} is the difference between the potentials of the gate and the source. If the drain current should increase for some reason, the voltage across R_3 will increase and this will make the gate potential more negative relative to the source potential. V_{gs} will be more negative and the drain current will fall, tending to compensate for the original increase.

The gate of an n-channel enhancement-mode mosfet must be held positive relative to the source and so a different bias method is necessary. Fig. 6.4a shows the usual circuit. Since zero current flows in the resistor R_1, the gate voltage is equal to $R_2/(R_1 + R_2)$ times the drain voltage of the fet. Should the requirement be for the gate voltage to be equal to the drain voltage, then R_2 can be omitted (Fig. 6.4b).

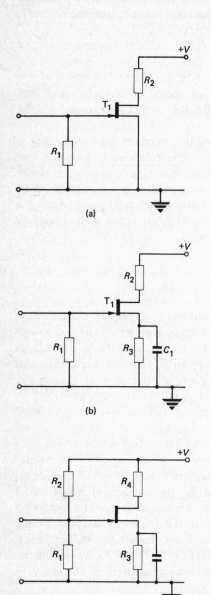

(a)

(b)

(c)

Fig. 6.3 Methods of biasing a jfet

Transistor Bias

The input terminals of a transistor can be biased to give a required operating point by the use of a suitable battery connected as shown in Fig. 3.5 or Fig. 3.7. It is more convenient, however, to derive the input bias current from the collector supply voltage.

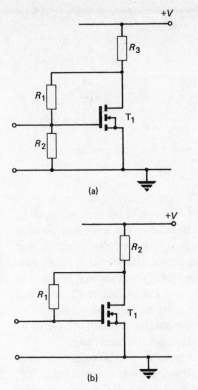

Fig. 6.4 Bias for an enhancement-mode mosfet

(1) The simplest method of establishing the operating point of a transistor connected in the common-emitter configuration is shown in Fig. 6.5. The potential difference across a forward-biased p-n junction is small and hence the base bias current I_b is given by

$$I_b \simeq \frac{E_{cc}}{R_b} \qquad (6.1)$$

The required bias current can be obtained by suitable choice of resistor R_b but, since a transistor bias arrangement is also required to stabilize the circuit against changes in the d.c. collector current, this simple circuit is not very effective and is rarely used.

The d.c. collector current may increase because an increase in the temperature of the collector-base junction will increase both the leakage current I_{CEO} and the short-circuit current gain h_{fe} of the transistor. In addition, a circuit is designed using the nominal h_{fe} of the transistor but individual transistors of the type used may have values of h_{fe} lying somewhere in between quoted maximum and minimum values. Typically, R_b would be many hundreds of kilohms and R_L about 3.3 kΩ.

The d.c. collector current is the sum of the amplified base bias current and the leakage current I_{CEO}, i.e.

$$I_c = h_{fe}I_b + I_{CEO} \qquad (6.2)$$

I_{CEO} is related to the collector leakage current when in the common-base configuration by the relation

$$I_{CEO} = I_{CBO}(1 + h_{fe})$$

EXAMPLE 6.1

A transistor whose h_{fe} may lie anywhere in the range 125 to 500 when the collector current is 2 mA is to be used in a single-stage amplifier using fixed current bias. Assuming a supply voltage of -6 V, the leakage current to be negligible, and the transistor to have the mean of its possible h_{fe} values, determine the base bias resistance needed. Find also the collector current which would then flow if the transistor has (i) minimum h_{fe} and (ii) maximum h_{fe}.

Solution
From equation (6.2), if I_{CEO} is negligible, then $I_b = I_c/h_{fe}$.
The mean value of h_{fe} is $625/2 \simeq 312$ and so the required base bias current is

$$\frac{2 \times 10^{-3}}{312} = 6.4 \ \mu A$$

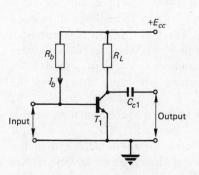

Fig. 6.5 Fixed current bias for a common-emitter connected transistor

Therefore, from equation (6.1), $R_b = \dfrac{6}{6.4 \times 10^{-6}} = 937.5 \ k\Omega$ (*Ans.*)

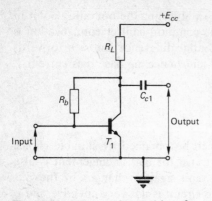

Fig. 6.6 Collector/base bias for a common-emitter connected transistor

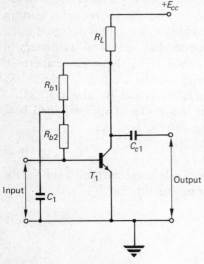

Fig. 6.7 Decoupling of collector/base bias circuit to prevent feedback

(i) When $h_{fe} = 125$, $I_c = 125 \times 6.4 \times 10^{-6} = 800\ \mu A$ (*Ans.*)

(ii) When $h_{fe} = 500$, $I_c = 500 \times 6.4 \times 10^{-6} = 3.2\ mA$ (*Ans.*)

(2) A better arrangement is shown in Fig. 6.6 and consists of connecting a bias resistor R_b between the collector and base terminals of the transistor. The base bias current I_b is given by the expression

$$I_b \simeq \frac{E_{cc} - R_L I_c}{R_b + R_L} \qquad (6.3)$$

where I_c is the steady collector current. The bias resistor R_b provides a path for the a.c. component of the collector current to feed into the base circuit, and if it is necessary to prevent this the a.c. component can be decoupled to earth as in Fig. 6.7. The bias resistor is replaced by two resistors giving the same total resistance and the decoupling capacitor C_1 is joined to their common point. It is necessary to split the resistance in this way because if C_1 were connected to the collector end of R_b, the output of the circuit would be short-circuited.

The circuit provides some degree of d.c. stabilization, the operation being, briefly, as follows. An increase in the d.c. collector current increases the voltage dropped across R_L and causes the collector/emitter voltage to fall and, since this voltage is effectively applied across the base resistor R_b, the base current falls also. The fall in base current results in a fall in the collector current which, to some extent, compensates for the original increase. The value of R_b would be a few hundred kilohms.

(3) For an improvement in the d.c. stabilzation the bias arrangement of Fig. 6.8 may be employed. The base of transistor T_1 is held at a positive potential V_b by the potential divider $(R_1 + R_2)$ connected across the collector supply, and the emitter is held at a positive potential V_e by the voltage developed across the emitter resistor R_3. The emitter/base bias potential is the difference between V_b and V_e and the component values chosen are such that the junction is forward-biased by a fraction of a volt. A bias current therefore flows into the emitter. Emitter resistor R_3 is adequately decoupled by capacitor C_1 to prevent an a.c. voltage at the signal frequency appearing across it and varying the emitter/base bias voltage. D.C. stabilization of the collector current is achieved in the following manner: an increase in the collector current is accompanied by an almost equal increase in emitter current. This results in an increase in the voltage V_e developed across the emitter resistor and this, in turn, reduces the forward bias of

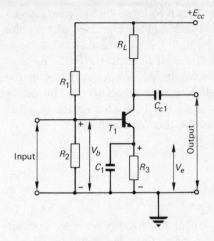

Fig. 6.8 Potential-divider bias for a common-emitter connected transistor

the emitter/base junction. The base current is reduced causing a decrease in the collector current that compensates for the original increase.

The POTENTIAL-DIVIDER circuit is far and away the most common bias and stabilization circuit, because good d.c. stabilization can be obtained with a wide variety of component values. Resistor R_1 is always of higher resistance value than resistor R_2; typically $R_1 = 100$ kΩ, $R_2 = 10$ kΩ or perhaps $R_1 = 47$ kΩ and $R_2 = 8.2$ kΩ. Resistance R_3 is generally about 1 kΩ and must be adequately decoupled by capacitor C_1; for this the value of C_1 must very large, generally at least 25 μF. The output coupling capacitor C_{c1} is provided as a d.c. blocking component to prevent current drawn from the collector supply passing into the output load. The value of C_{c1} is chosen so that the component will have negligible reactance at most of the frequencies of operation of the amplifiers. As the frequency of operation is reduced the reactance of C_{c1} will increase and increasingly some of the a.c. output voltage will be dropped across it and not across the load as required. This means that the voltage gain of the circuit falls with increase in frequency, and the value of C_{c1} is chosen to give the desired low-frequency response and is, typically, about 1 μF.

The gain at high frequencies will decrease with increase in frequency because of unavoidable circuit capacitance which effectively shunt the signal path. Further loss of gain will occur if the current gain of the transistor falls with increase in frequency, but this effect can be easily avoided by choosing a transistor with a sufficiently high f_t. The gain/frequency characteristic of an audio frequency amplifier is shown in Fig. 6.9. It can be seen that the gain is constant over a wide frequency band and falls at both low and high frequencies. The *bandwidth* of an amplifier is the band of frequencies over which the gain is not more than $1/\sqrt{2}$ times the maximum gain.

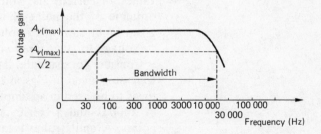

Fig. 6.9 Gain/frequency characteristic of an audio-frequency amplifier

Determination of Gain using a Load Line

The voltage gain of a fet amplifier or the current gain of a transistor circuit can be determined with the aid of a "load

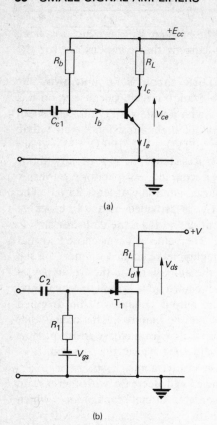

Fig. 6.10 Currents and voltages in basic amplifiers

line" drawn on the output current/output voltage characteristics of the device. The currents and voltages existing in the collector or drain circuit of a simple resistance-loaded amplifier are shown in Figs. 6.10a and b respectively. In each of these circuits the d.c. collector or drain current flows in the load resistor R_L and develops a voltage across it. The direct voltage which is applied across the transistor or fet is equal to the supply voltage minus the d.c. load voltage.

Thus, referring to Fig. 6.10,

$$V_{ce} = E_{cc} - I_c R_L \qquad (6.4)$$

$$V_d = V - I_d R_L \qquad (6.5)$$

Equations (6.4) and (6.5) are of the form $y = mx + c$ and are therefore equations to a straight line. In order to draw a straight line it is only necessary to plot two points; these points can best be determined in the following manner. Point A: Let $I_c = I_d = 0$ in equations (6.4) and (6.5) respectively, then,

$$V_{ce} = E_{cc} \quad \text{and} \quad V_d = V$$

Point B: Let $V_{ce} = V_d = 0$ in equations (6.4) and (6.5) respectively, then

$$0 = E_{cc} - I_c R_L \quad \text{or} \quad I_c = \frac{E_{cc}}{R_L}$$

$$0 = V - I_d R_L \quad \text{or} \quad I_d = \frac{V}{R_L}$$

If the points A and B are marked on the characteristics and then joined together by a straight line, the line drawn is the LOAD LINE for the particular values of load resistance and supply voltage. The load line can be used to determine the values of current and voltage in the output circuit, since the ordinate of the point of intersection of the load line and a given input current or voltage curve gives the output current or voltage for that input signal.

Consider for example, the output characteristics of an n-p-n transistor given in Figs. 6.11a and b, and suppose the transistor is to be used in an amplifier with a collector load resistance of 2000 Ω and a collector supply voltage of 10 V. The two points, A and B, that locate the ends of the load line are at $I_c = 0$, $V_{ce} = E_{cc} = 10$ V, and at $V_{ce} = 0$, $I_c = E_{cc}/R_L = 10/2000 = 5$ mA. These points have been located on the characteristics and the load line drawn between them. The operating point should be chosen to lie approximately in the middle of the load line, and has been selected as the point marked P. The required base bias current is then equal to 20 μA. The d.c. collector/emitter voltage ($V_{ce} = E_{cc} - I_c R_L$) is found by projecting vertically downwards from the operating point to the voltage axis. This step is shown by a dotted line in Fig. 6.11a and it determines the standing (or quiescent) collector/emitter voltage as 5.6 V. Similarly, the d.c. collector current which flows is found by projecting horizontally from the operating point towards the current axis. Thus, the d.c. collector current is equal to 2.2 mA.

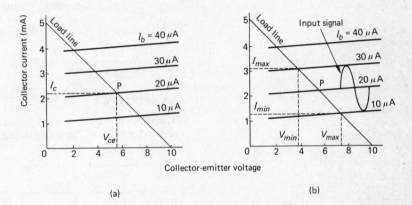

Fig. 6.11 (a) (b)

The d.c. power taken from the collector supply is given by the product of the collector supply voltage and the d.c. collector current, i.e. $10 \times 2.2 \times 10^{-3} = 22$ mW. The d.c. power P_c dissipated at the collector of the transistor is the difference between the d.c. power supplied to the circuit and the d.c. power dissipated in the load resistance. Hence,

$$P_c = 22 \times 10^{-3} - (2.2 \times 10^{-3})^2 \times 2000 = 12.32 \text{ mW}$$

P_c is also equal to the product of the d.c. collector current and the quiescent collector/emitter voltage; thus,

$$P_c = 2.2 \times 10^{-3} \times 5.6 = 12.32 \text{ mW}$$

The load line can also be used to find the variation in collector current and collector/emitter voltage which is produced by the application of a signal to the base of the transistor. Suppose as in Fig. 6.11b that a sinusoidal signal of peak value 10 μA is applied. This signal is superimposed upon the base bias current of 20 μA and so the base current is varied from a minimum value of 10 μA to a maximum value of 30 μA. The corresponding values of collector current and collector/emitter voltage are determined by projecting to the current and voltage axes from the points of intersection of the load line and the curves for $I_b = 10 \mu$A and $I_b = 30 \mu$A. This has been shown by the dotted lines drawn on Fig. 6.11b. The collector current is varied from a minimum value $I_{min} = 1.3$ mA to a maximum value $I_{max} = 3.25$ mA. The collector-emitter voltage is varied from a minimum value $V_{min} = 3.6$ V to a maximum value $V_{max} = 7.3$ V. The a.c. component of the collector current has a peak-to-peak value of $(3.25 - 1.3)$ mA or 1.95 mA, while the peak-to-peak value of the a.c. component of the collector/emitter voltage is $(7.3 - 3.6)$ V or 3.7 V. The current gain A_i of the amplifier is

$$A_i = \frac{\text{Peak-to-peak change in collector current}}{\text{Peak-to-peak change in base current}}$$

Hence,

$$A_i = \frac{1.95 \times 10^{-3}}{20 \times 10^{-6}} = 97.5$$

EXAMPLE 6.2

The transistor used in a single-stage audio-frequency amplifier with a resistance load of 2000 Ω has the data given in the Table.

V_{ce}(V)	I_c(mA)			
	$I_b = 20\ \mu A$	$I_b = 40\ \mu A$	$I_b = 60\ \mu A$	$I_b = 80\ \mu A$
2	0.85	1.55	2.32	3.08
4	1.00	1.74	2.56	3.35
6	1.13	1.92	2.76	3.60
8	1.30	2.13	3.00	3.85

Plot the output characteristics of the transistor and draw the load line assuming a collector supply voltage E_{cc} of 8 V.

(i) Select a suitable operating point.
(ii) Determine the current gain A_i when an input signal producing a base current swing of 20 μA about the chosen bias current is applied to the circuit.
(iii) Assuming the input resistance of the transistor is 1200 Ω determine the voltage gain A_v.
(iv) Calculate the power gain A_p.

Solution
The output characteristics are shown plotted in Fig. 6.12. The d.c. load line must be drawn between the points

$$I_c = 0, \quad V_{ce} = E_{cc} = 8 \text{ V}, \quad \text{and}$$
$$V_{ce} = 0, \quad I_c = E_{cc}/R_L = 8/2000 = 4 \text{ mA}$$

(i) Since the input signal has a peak value of $\pm 20\ \mu$A a suitable base bias current is 40 μA, the operating point is then P.
(ii) When a signal of $\pm 20\ \mu$A peak is applied to the transistor, the base current varies between 20 μA and 60 μA. Projection from the intersection of the load line and the 20 μA and 60 μA base current curves to the current axes gives the resulting values of collector current as 1.15 mA and 2.45 mA.

The peak-peak collector current swing is therefore $2.45 - 1.15$ or 1.30 mA and the current gain is

$$A_i = \frac{1.3 \times 10^{-3}}{40 \times 10^{-6}} = 32.5 \qquad (Ans.)$$

(iii) If the input resistance of the transistor is $1200\,\Omega$ the a.c. voltage applied to the transistor must be

$$\pm 20 \times 10^{-6} \times 1200 \quad \text{or} \quad \pm 24 \times 10^{-3}\,\text{V}$$

Projecting from the intersection of the load line and the appropriate base current curves to the voltage axis gives the peak-peak collector voltage as

$$5.7 - 3.08 = 2.62\,\text{V}$$

Voltage gain $A_v = \dfrac{2.62}{48 \times 10^{-3}} = 54.59 \qquad (Ans.)$

Alternatively, using equation (3.6),

$$A_v = \frac{A_i R_L}{R_{IN}} = \frac{32.5 \times 2000}{1200} = 54.17 \qquad (Ans.)$$

(iv) The power output of the transistor is the product of the r.m.s. values of the a.c. components of the collector current and the collector-emitter voltage. Therefore

$$\text{Output power} = \frac{\text{peak-peak } I_c}{2\sqrt{2}} \times \frac{\text{peak-peak } V_{ce}}{2\sqrt{2}}$$

$$= \tfrac{1}{8}[(I_{c(max)} - I_{c(min)})(V_{ce(max)} - V_{ce(min)})]$$

$$= \tfrac{1}{8}(1.3 \times 10^{-3} \times 2.62) = 4.26 \times 10^{-4}\,\text{W} \qquad (Ans.)$$

The input power delivered to the transistor is

$$I_{b(rms)}^2 R_{IN} = \left(\frac{20 \times 10^{-6}}{\sqrt{2}}\right)^2 \times 1200 = 2.4 \times 10^{-7}\,\text{W}$$

Power gain $A_p = P_{out}/P_{in} = 4.26 \times 10^{-4}/2.4 \times 10^{-7} = 1775 \qquad (Ans.)$

Alternatively using equation (3.7)

$$A_p = A_i^2 R_L/R_{IN} = A_i A_v = 32.5 \times 54.17 = 1761 \qquad (Ans.)$$

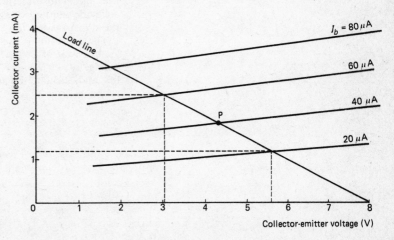

Fig. 6.12

EXAMPLE 6.3

Determine the mutual conductance of the transistor in Example 6.2. Use this value to calculate the voltage gain of the circuit.

Solution
From equation (3.17)

$$g_m = A_i/R_{IN} = 32.5/1200 = 27 \text{ mS} \quad (Ans.)$$

Alternatively, $g_m = 40$ mS per mA collector current.

$$\delta I_c = 1.30/2 = 0.65 \text{ mA}$$

Hence $g_m = 0.65 \times 40 = 26$ mS

$$\text{Voltage gain} = g_m R_L = 27 \times 10^{-3} \times 2 \times 10^3 = 54 \quad (Ans.)$$

Exercises

6.1. With the aid of sketches describe any one type of transistor, and explain briefly why it can act as an amplifier.
Give a circuit of a single-stage audio-amplifier using a transistor.
What effect has temperature change on the operation of such an amplifier? (C&G)

6.2. The data given in Table A refer to a transistor in the common-emitter configuration.

Table A

	Collector current (mA)		
Collector/emitter voltage (V)	Base current −40 μA	Base current −60 μA	Base current −80 μA
−3	−1.6	−1.8	−2.0
−5	−1.75	−2.1	−2.5
−7	−1.9	−2.4	−2.95
−9	−2.05	−2.7	−3.4

Plot the collector current/collector voltage characteristics and use these to determine: (*a*) the current gain when the collector/emitter voltage is −6 V, (*b*) the output resistance of the transistor for $I_b = -60 \mu$A.
The transistor is used as a common-emitter amplifier with a load resistor of 2000 Ω and is operated with a collector supply voltage of −9 V and a base current of −60 μA. Draw the load line and calculate the actual current gain when the base current is changed by 20 μA during each half-cycle. (C&G)

6.3. The data given in Table B refer to a transistor in the common-emitter configuration.

Table B

Collector/ emitter voltage (V)	Collector current (mA)			
	Base current −20 μA	Base current −60 μA	Base current −100 μA	Base current −140 μA
−1	−0.8	−2.8	−4.8	−6.8
−3	−1.2	−3.3	−5.4	−7.5
−5	−1.6	−3.8	−6.0	−8.2
−7	−2.0	−4.3	−6.6	−8.9
−9	−2.4	−4.8	−7.2	−9.6

Plot the collector current/collector voltage characteristics for base currents of −20, −60, −80 and −100 μA. Use the characteristics to determine: (a) the output resistance for $I_b = -100$ μA, and (b) the current gain for $V_{ce} = -6$ V. The transistor is to be used in a common-emitter amplifier circuit with a load of 1000 Ω and a collector supply voltage of −10 V. Draw the load line. (C&G)

6.4. (a) Sketch the circuit diagram of a single stage Class A amplifier in which the active device is a transistor. (b) Describe how the operating point will alter with variation in the temperature of the device. (c) Explain the principle of operation of the particular type of transistor used in your diagram.

6.5. (a) The data given in Table C refer to a transistor in the common-emitter configuration.

Table C

Collector/ emitter voltage (V)	Collector current (mA)		
	Base current 0 μA	Base current −20 μA	Base current −40 μA
−3	−0.20	−0.91	−1.6
−5	−0.23	−0.93	−1.7
−7	−0.26	−0.97	−1.85
−9	−0.30	−1.00	−2.05

Plot the collector current/collector voltage characteristics for base currents of 0, −20, and −40 μA. (b) Use these characteristics to determine (i) the current gain when the collector voltage is −6 V, (ii) the output resistance for a base current of −40 μA. (c) Draw the load line for a collector resistance of 4000 Ω and a supply voltage of −10 V. (d) Determine the power dissipated at the collector when the base current is −20 μA. (C&G)

6.6. The data given in Table D refer to a transistor in the common-emitter configuration.

Table D

Collector/emitter voltage (V)	Collector current (mA)			
	Base current $-20\,\mu A$	Base current $-40\,\mu A$	Base current $-60\,\mu A$	Base current $-80\,\mu A$
-3	-0.91	-1.6	-2.3	-3.0
-5	-0.93	-1.7	-2.5	-3.25
-7	-0.97	-1.85	-2.7	-4.05

Plot the collector current/collector voltage characteristics for base currents of -20, -40, -60, and $-80\,\mu A$ and use these curves to determine: (a) the current gain when the collector voltage is $-6\,V$, (b) the output resistance of the transistor when $I_b = -40\,\mu A$.

The transistor is to be used in a common-emitter amplifier with a load resistance of $2500\,\Omega$ and a collector supply voltage of $-10\,V$. Draw the load line and use this to find the collector current for a collector voltage of $-5\,V$.   (C&G)

6.7. Draw the circuit diagram of a small-signal common-emitter amplifier and explain the operation of the bias circuit for the provision of (i) the required operating point and (ii) d.c. stability. Why is d.c. stability necessary? What is the phase relationship between the input and output voltage?

6.8. A transistor has the data given in Table E.

Table E

Collector/emitter voltage (V)	Collector current (mA)			
	$I_b = 20\,\mu A$	$I_b = 40\,\mu A$	$I_b = 60\,\mu A$	$I_b = 80\,\mu A$
2	0.9	1.55	2.2	2.85
4	0.92	1.65	2.4	3.05
6	0.95	1.77	2.55	3.25
8	0.98	1.90	2.75	3.50

Plot the output characteristics and determine from them the short-circuit current gain and the output resistance of the transistor.

The transistor is used in an amplifier circuit with a collector load resistance of $2200\,\Omega$, base bias current of $50\,\mu A$, and a collector supply voltage of $6\,V$. Draw the load line and mark the operating point. If the transistor has an input resistance of $1000\,\Omega$ determine the current, voltage and power gains of the amplifier when a sinusoidal signal of $30\,\mu A$ peak is applied to the base.

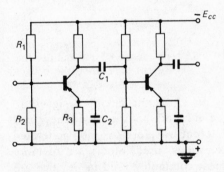

Fig. 6.13

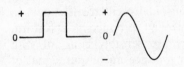

Fig. 6.14

6.9. Draw circuit diagrams showing how a transistor may be used in a single-stage amplifier with (*a*) common base and (*b*) common emitter. What is the phase relationship between the input and output voltage waveforms in each case when a sinusoidal input signal is applied?

Give some reasons why the common-emitter configuration is normally chosen.

6.10. Draw the circuit diagram of a common-emitter transistor amplifier employing potential-divider bias. State why d.c. stabilization of the operating point is necessary and explain how your bias circuit operates to provide such stability.

6.11. (*a*) Explain with the aid of diagrams the following amplifier classifications (fet or transistor examples may be used): (i) Class A, (ii) Class B, (iii) Class C. (*b*) With reference to transistor amplifiers: (i) what is thermal runaway, (ii) how can thermal stability be improved? (C&G)

6.12. With reference to the small-signal amplifier of Fig. 6.13, (*a*) Suggest values for R_1 and R_2 and state their purpose, (*b*) Suggest values for C_1, C_2 and R_3, (*c*) Describe the functions R_3, C_2 and C_3, (*d*) State the purpose of C_1, (*e*) What would be the effect on the performance if C_1 was much reduced in value? (part C&G)

Short Exercises

6.13. Draw the circuit diagram of a single-stage amplifier with a collector load resistor R_L and fixed current bias. If the collector supply voltage is $-24\,\text{V}$, the load resistance is $4700\,\Omega$ and the transistor is biased to have a collector current of $2.5\,\text{mA}$, calculate the voltage V_{ce} across the transistor.

6.14. Explain why it is necessary to bias a transistor to have a particular operating point.

6.15. The waveforms shown in Fig. 6.14 are applied in turn to the input terminals of a common-emitter transistor amplifier. Sketch the expected output waveforms.

6.16. What is meant by *thermal runaway* of a transistor and why are heat sinks often employed?

6.17. A transistor is connected in an amplifier with a load resistance of $1500\,\Omega$ and then has a current gain of 120 and an input resistance of $1100\,\Omega$.

(i) Is the transistor connected in the common-base or in the common-emitter configuration?

(ii) Is the short-circuit current gain of the transistor greater than, or less than, 120?

(iii) Calculate the voltage gain of the amplifier.

6.18. Draw sketches to illustrate the meanings of the terms Class A, Class B and Class C bias of an amplifying device.

6.19. Draw diagrams to illustrate the three basic transistor amplifier configurations. Compare their voltage gains, input impedances, and output impedances.

6.20. What is meant by the bandwidth of an amplifier? List some typical bandwidths used in practice.

6.21. By applying Kirchhoff's voltage law to the circuit of Fig. 6.6 derive equation 6.3 (p. 86). Assume V_{be} is negligible.

7 Waveform Generators

Waveforms

A waveform generator is an electronic circuit designed to produce an alternating e.m.f. of known frequency and waveform. A variety of different waveforms can be generated by these circuits, some are widely used, others are of more limited application. The most commonly used in practice are the sinusoidal, the rectangular, and the sawtooth waveforms. Sinusoidal waveforms are produced by waveform generators known as OSCILLATORS, while both rectangular and sawtooth waveforms can be generated by ASTABLE MULTIVIBRATORS or by BLOCKING OSCILLATORS.

The output waveform of a sinusoidal oscillator is shown in Fig. 7.1. Oscillators are designed to produce a waveform whose amplitude and frequency are sensibly constant with time; some circuits are made to produce an output signal of fixed amplitude and/or frequency, while other circuits have their output amplitude and/or frequency continuously variable.

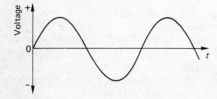

Fig. 7.1 Sinusoidal waveform

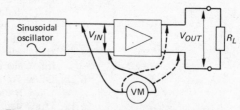

Fig. 7.2 Measurement of amplifier gain

SINUSOIDAL WAVEFORMS are used for many varied purposes in electronic/telecommunication circuitry. Many measurement techniques are based on the use of a sinusoidal signal of known characteristics. For example, the gain of an amplifier can be measured using the arrangement shown in Fig. 7.2. The amplitude and frequency of the output of the oscillator are set to appropriate values and the voltage V_{OUT} developed across the load resistance is measured. The voltage gain of the amplifier is then given by the ratio V_{OUT}/V_{IN}. The input voltage V_{IN} must be adjusted to a value low enough to ensure that the amplifier is not driven into operation on the non-linear parts of its output characteristics, otherwise distortion of the output waveform and a reduction in gain will occur. It is good practice to use the same voltmeter to measure both the input and output voltages since errors due to voltmeter inaccuracies are then minimized.

Sinusoidal waveforms are employed as *carriers* of information signals in both radio and line systems; the information signal is made to vary either the amplitude or the frequency of the *carrier wave*. The process, known as *modulation*, results in the information signal being shifted to another part of the frequency spectrum. A further common application of the sinusoidal waveform is to be found in radio and television receivers; here the sinusoidal output of a *local oscillator* is applied to a circuit, known as a mixer, together with the signal to which the receiver has been tuned. A *mixing* process takes place in which the wanted signal is shifted to a constant *intermediate frequency* that the receiver has been designed to handle.

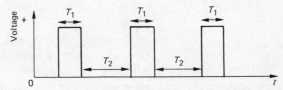

Fig. 7.3 Rectangular waveform

Fig. 7.3 shows a RECTANGULAR WAVEFORM. The waveform has been drawn as varying between zero volts and a positive voltage, but the inverse, i.e. between zero volts and a negative value, is equally common. The MARK/SPACE RATIO of the waveform is T_1/T_2 and the DUTY CYCLE is $T_1/(T_1+T_2)$. The duty cycle may be a fixed quantity or, with some waveform generators, it may be adjustable. The periodic time of the waveform is (T_1+T_2) and the *pulse repetition frequency* (p.r.f.) is the number of pulses occurring per second and is equal to the reciprocal of the periodic time. If T_1 is adjusted to be equal to T_2 then a square waveform is obtained.

A rectangular waveform can be used to test the performance of an electronic system. If, for example, a rectangular waveform is applied to the input of an amplifier, any change in the waveform at the output of the circuit will indicate the extent to which the amplifier performance is inadequate. If the gain of the amplifier is not constant over a wide enough bandwidth, the output pulses will not be rectangular. Should the gain start to decrease at too high a frequency, the output

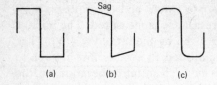

Fig. 7.4 Sag on a rectangular waveform

pulse waveform will exhibit *sag*, as shown in Fig. 7.4b. The high-frequency end of the gain/frequency characteristic of an amplifier should remain constant up to a frequency sufficiently high to ensure that the lower harmonics contained in the input rectangular waveform are passed. If the high-frequency response is not good enough the output pulses will be rounded off as shown by Fig. 7.4c. Square waveforms can also be modulated by an information signal; it is possible to make the amplitude, or the time duration, or the position in time, of the pulses vary in accordance with the characteristics of the modulating signal.

Other applications of rectangular waveforms are

(a) Synchronizing pulses are added to the picture signal produced by a television camera to ensure that the scanning of the television receiver screen is kept in step with the scanning of the original picture by the television camera.

(b) Rectangular pulses, generally known as CLOCK PULSES, are often used in digital electronic systems to ensure that various sub-systems operate at the correct instants in time.

The SAWTOOTH WAVEFORM (Fig. 5.12) consists of a voltage that rises linearly with time, known as a RAMP, which when it reaches its maximum value falls rapidly to zero. Immediately the voltage has fallen to zero it begins another ramp and so on. Sawtooth waveforms are employed whenever a voltage or a current that increases at a constant rate is required. Examples of the use of such waveforms are the timebase voltage applied across the X-plates of the cathode ray tube in an oscilloscope, and the currents passed through the line and field scan coils of a television receiver tube. Sawtooth waveforms are also used in some kinds of digital voltmeter in which the number of rectangular pulses passing through an electronic gate in the time the ramp takes to reach the same value as the voltage under test are counted to give a measure of the voltage.

In this book only oscillators providing a sinusoidal output waveform are to be discussed. Further, discussion will be limited to the simple types of oscillator in which the frequency of oscillation is determined by a parallel-tuned *L-C* circuit.

L-C Oscillators

A sinusoidal oscillator (fet or transistor) can be regarded as an amplifier that provides its own input signal, this input signal being derived from the output signal (see Fig. 7.5). This is possible because the signal level required at the input termi-

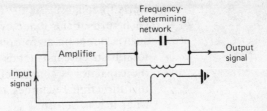

Fig. 7.5 The principle of an oscillator

nals of an amplifier is considerably less than the level of the amplified output signal. The transistor or fet acts as a convertor of electrical energy, taking d.c. power from the supply and converting a part of it into a.c. power in the output signal.

Oscillators are widely used in electronic, radio, and line communication equipment as sources of alternating e.m.f. and are generally required to be extremely stable in frequency. Frequency stability of the *highest* order cannot be obtained with the simple oscillators to be described and other high-stability oscillators are used in practice. Other characteristics of importance are the purity of the output waveform, and the constancy of the output level with changes in frequency and/or in power supply voltage.

The Oscillatory Circuit

If a capacitor, C farads, is charged from a d.c. source, a p.d. V volts will be developed across its terminals and an amount of electric energy, $\frac{1}{2}CV^2$ joules, will be stored in its dielectric. Consider such a charged capacitor to be connected across an inductor as shown in Fig. 7.6. A complete circuit exists and so the capacitor will discharge through the inductor and a current i will flow. This current begins to flow the instant the capacitor is connected across the inductor, and it rises rapidly to a maximum value when the capacitor has fully discharged and there is zero voltage across its plates. The flow of a current in a conductor produces a magnetic field around that conductor; associated with the current flow in the inductor, therefore, is a magnetic field that reaches its maximum value at the same times as does the current. The energy stored in the magnetic field at this time is equal to $\frac{1}{2}LI^2$ joules, where L is the inductance of the inductor in henrys and I is the maximum value of the current in amperes. All the energy originally stored in the capacitor has now disappeared (since $V = 0$) and has been partly converted into magnetic energy and partly lost as power dissipation in the resistance r of the circuit.

Since the p.d. across the capacitor terminals is now zero, the current starts to fall and the magnetic field about the inductor

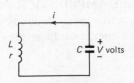

Fig. 7.6 The oscillatory circuit, initial conditions

starts to collapse. As the field collapses, an e.m.f. is induced in each turn of the inductor that, according to Lenz's law, is in such a direction as to oppose the force creating it; that is, the total induced e.m.f. tends to keep the current flowing. Because the capacitor is completely discharged, further current flow in this direction recharges it—but with the polarity opposite to what it was before.

When the magnetic field has completely collapsed, the current has fallen to zero and the capacitor is fully re-charged to a voltage somewhat less than before, say $(V - \delta V)$, where δV is a small voltage increment (Fig. 7.7). Almost all of the magnetic energy has been converted back to the form of electric energy stored in the dielectric of the capacitor, some energy having again been lost as $i^2 r$ dissipation in the circuit resistance. The capacitor now starts to discharge through the inductor again but this time the current flow is in a direction opposite to what it was before. A magnetic field is again set up around the inductor that increases with increase in the discharge current. When the capacitor is fully discharged the current starts to fall and the collapsing magnetic field induces an e.m.f. in the inductor windings that tends to maintain the current in its new direction. The capacitor is recharged with its original polarity by this current and when it is fully charged (to a voltage less than before) one cycle of the oscillatory current has been completed (see Fig. 7.8)

A continual interchange of energy between the capacitor and the inductor takes place at a constant frequency, but with the amplitude of the oscillatory current decreasing steadily until, eventually, the oscillation dies away. An oscillation of this type, shown in Fig. 7.9, is known as a DAMPED OSCILLATION. The rate at which the oscillation dies away depends upon the circuit resistance; the greater the resistance the sooner the oscillations disappear.

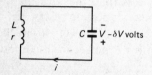

Fig. 7.7 The oscillatory circuit, conditions after the first half-cycle of oscillation

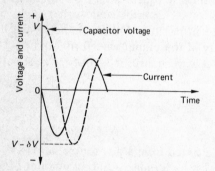

Fig. 7.8 A cycle of oscillation

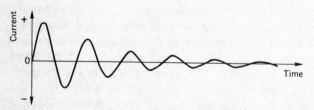

Fig. 7.9 A damped oscillation

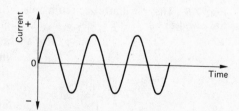

Fig. 7.10 An undamped oscillation

If energy can be supplied to the oscillatory circuit to replace the energy lost by $i^2 r$ dissipation, an undamped oscillation can be obtained. An UNDAMPED oscillation is shown in Fig.

7.10, and it is clear that the current amplitude is constant and the oscillation does not die away but can be maintained indefinitely. The energy supplied to the oscillatory circuit must be sufficiently large to make good the losses and must be in phase with the oscillation.

If the losses in the tuned circuit are low the frequency f_0 of oscillation is given approximately by the expression

$$f_0 \simeq \frac{1}{2\pi\sqrt{(LC)}} \text{ Hz} \qquad (7.1)$$

where L is in henrys and C is in farads.

The energy that must be supplied to the oscillatory circuit to maintain oscillations is provided by the amplifier section of the oscillator (Fig. 7.5). The requirements that must be satisfied for an oscillator to oscillate are: first, the loop phase shift must be zero, and secondly the gain around the loop must be unity. The first requirement is to ensure that the energy supplied by the amplifier to overcome the losses in the circuit is in phase with the oscillatory circuit current (this is known as *positive feedback*), and the second requirement is necessary because if the loop gain were to be less than unity the circuit losses would not be fully compensated for and the oscillations would gradually die away.

When an oscillator is first switched on, a current surge in the frequency-determining network produces a voltage, at the required frequency of operation, across the network. A fraction of this voltage is fed back to the input terminals of the amplifier and is amplified to reappear across the network in phase with the original voltage. A fraction of this larger voltage is then fed, in turn, back to the input and is further amplified and so on. In this way the amplitude of the signal voltage builds up until it is limited in some way. Once the oscillations have built up to the required amplitude, the loop gain is reduced to unity. The gain may be reduced either by the valve or transistor being driven into saturation or by operating the circuit under Class C conditions.

Types of L-C Oscillator

The bias and d.c. stabilization requirements of a Class A oscillator are similar to those used for an amplifier and the same circuitry is used. Oscillators may also be operated under Class C conditions and then generally incorporate leaky base bias, the circuit for which is given in Fig. 7.11.

When a transistor has zero bias voltage it is non-conducting and so Class A bias must be provided to ensure that the transistor will conduct immediately the power supplies are switched on, otherwise the circuit would not oscillate. Generally, the required initial bias is provided by a potential divider as in the circuit shown in Fig. 7.11.

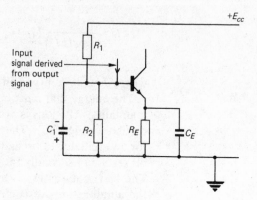

Fig. 7.11 The leaky-base bias circuit

Some oscillator circuits use Class A bias only and rely on the build-up of oscillations driving the transistor into saturation and thereby reducing the loop gain to unity. Other circuits employ **LEAKY-BASE** Class C bias. This is obtained by connecting a capacitor of suitable value across the lower of the two potential divider resistors; for example capacitor C_1 in Fig. 7.11.

The operation of the leaky-base circuit is as follows. As the amplitude of the oscillations builds up, the point is reached where the amplitude of the negative half-cycles of the input signal at the base is larger than the positive voltage supplied by the potential divider. The capacitor is then charged with the polarity shown which biases the transistor beyond the cut-off point. During the next positive half-cycle of the input signal, the capacitor begins to discharge at a rate such that it has not fully discharged before the next negative half-cycle arrives. During the next negative half-cycle, the charge lost by the capacitor is replaced, plus some extra charge. In each complete cycle of the input signal the capacitor receives more charge than it loses and so the voltage across its terminals builds up. Eventually, an equilibrium condition is reached in which the charge gained in each input signal cycle is equal to the charge lost and the capacitor voltage settles down to a constant value. This constant value, which is the bias voltage applied to the transistor, is proportional to the amplitude of the input signal.

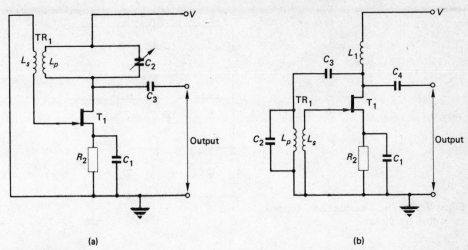

(a)   (b)

Fig. 7.12 Tuned-drain oscillators: (*a*) series-fed and (*b*) parallel-fed

The Tuned-drain Oscillator

A tuned-drain oscillator is one in which the frequency-determining tuned circuit is connected in the drain circuit of the fet. The supply voltage may be fed to the drain either in series with the tuned circuit or in parallel with the tuned circuit. The first arrangement, known as SERIES FEED, is shown in Fig. 7.12*a* and the second arrangement, known as PARALLEL FEED, is shown in Fig. 7.12*b* The parallel-feed circuit has the advantage that the tuned circuit is at earth potential and not supply potential. In both circuits the frequency of oscillation is determined by the product L_pC_2 and bias is provided by R_1 and R_2. In Fig. 7.12*b* the inductor L_1 is an r.f. choke whose function is to block r.f. currents from the supply line whilst dropping the minimum d.c. voltage. Capacitors C_3 and C_4 are d.c. blocking capacitors that have negligible reactance at the oscillation frequency and that block the supply voltage.

The action of either of these circuits is as follows. When the supply voltage is switched on, any noise or small fluctuation in the gate circuit is amplified and causes an oscillatory current to be set up in the tuned circuit. The oscillatory current in the primary winding L_p of the transformer induces an e.m.f. at the same frequency into the secondary winding L_s and this voltage is applied to the gate of the fet. The fet introduces 180° phase shift and the transformer connections must be arranged to give a further 180° phase change to achieve an overall phase change around the oscillator loop of zero. Also, the coupling between the transformer windings must be tight enough to ensure that the loop gain is greater than unity; otherwise the oscillations will die away.

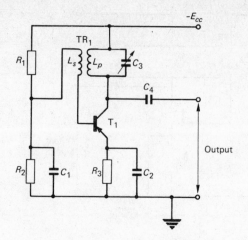

Fig. 7.13 The series-fed tuned-collector oscillator

The Tuned-collector Oscillator

The circuit of a tuned-collector oscillator is given in Fig. 7.13. R_1, R_2, R_3 and C_2 are Class A bias and d.c. stabilization components, TR_1 is an r.f. transformer, C_3 a variable tuning capacitor, and C_4 a blocking capacitor. C_1 and R_2 provide Class C bias. The operation of the circuit is similar to the operation of the tuned-drain oscillator previously described.

The frequency of oscillation f_0 is given approximately by

$$f_0 \simeq \frac{1}{2\pi\sqrt{(L_p C_3)}} \text{ Hz} \tag{7.2}$$

The output signal can be taken off as shown in the diagram, or, alternatively, taken from a third winding L_0 coupled to L_p and L_s as shown in the parallel-feed circuit of Fig. 7.14a. Finally, the oscillatory voltage may be fed into the emitter circuit instead of the base (Fig. 7.14b).

Fig. 7.14 Tuned-collector oscillators:
(a) parallel-fed and
(b) employing emitter coupling

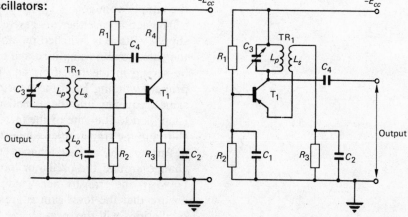

(a)   (b)

EXAMPLE 7.1

A tuned-collector oscillator has a fixed inductance of 75 μH and is to be tunable over the frequency band 500 kHz to 1500 kHz. Calculate the range of the variable capacitor to be used.

Solution
From equation (7.2),

$$f_0^2 = \frac{1}{4\pi^2 L_p C_3}$$

Transposing

$$C_3 = \frac{1}{4\pi^2 f_0^2 L_p}$$

When $f_0 = 500$ kHz,

$$C_3 = 1/(4\pi^2 500^2 \times 10^6 \times 75 \times 10^{-6}) = 1351 \text{ pF}$$

When $f_0 = 1500$ kHz,

$$C_3 = 1351/9 = 150 \text{ pF}$$

Capacitor range required is 150–1351 pF   (*Ans.*)

EXAMPLE 7.2

The frequency-determining circuit of a tuned-collector oscillator resonates at 6 MHz. If the value of the capacitance is increased by 55%, what is the new resonant frequency?.

Solution
From equation (7.2)

$$6 \times 10^6 = \frac{1}{2\pi\sqrt{(LC)}} \tag{7.3}$$

and

$$f_0 = \frac{1}{2\pi\sqrt{(L \times 1.55C)}} \tag{7.4}$$

Dividing equation (7.4) by (7.3),

$$\frac{f_0}{6 \times 10^6} = \frac{1}{\sqrt{1.55}}$$

Therefore

$$f_0 = \frac{6 \times 10^6}{\sqrt{1.55}} = 4.819 \text{ MHz} \quad (\textit{Ans.})$$

The Tuned-base Oscillator

The circuit of a tuned-base transistor oscillator is shown in Fig. 7.15. Conventional Class A bias and d.c. stabilization circuitry is used and the frequency of oscillation is determined by the parallel-tuned circuit of L_s and C_3 (i.e. $f_0 = 1/2\pi\sqrt{(L_sC_3)}$). Feedback from output to input is via capacitor C_1 and the r.f. transformer TR_1. Capacitor C_4 is necessary to prevent the bias for transistor T_1 being determined primarily by the low d.c. resistance of winding L_s. Parallel feed of the collector supply voltage is shown and has the usual advantage over series feeding in that no d.c. current flows in the windings of the transformer. The operation of the circuit is similar to the operation of the preceding circuits.

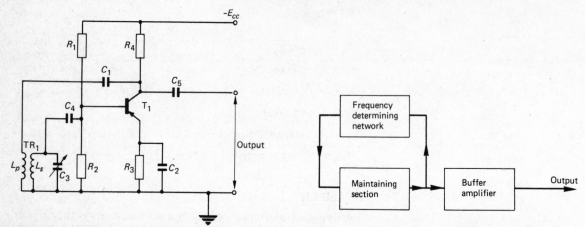

Fig. 7.15 The tuned-base oscillator

Fig. 7.16 The use of a buffer amplifier

Frequency Stability

The frequency of oscillation of an oscillator is primarily dependent upon the resonant frequency of its tuned circuit but it is also a function of various other parameters. The main causes of frequency instability are changes in the value of the tuned-circuit capacitance and inductance because of changes in temperature, changes in fet or transistor parameters due to fluctuations in power supplies, and variations in the external load.

The temperature cause of frequency instability can be minimized by using inductors and capacitors in the tuned circuit that have small temperature coefficients, and keeping the tuned circuit at as constant a temperature as possible. Fluctuations in the power supplies can be reduced by the use of adequate power stabilization circuitry, and variations in the external load can effectively be removed by feeding the load via a buffer amplifier. Fig. 7.16 shows the use of a buffer amplifier in an oscillator. The amplifier is an ordinary audio-frequency or radio-frequency amplifier—depending upon the oscillation frequency—whose functions are to isolate the oscillator from any changes in the load and to increase the output power level.

When very good frequency stability is required, none of these measures is adequate and crystal oscillators are used. A crystal oscillator is an oscillator whose frequency is determined by a piezoelectric crystal such as quartz.

Exercises

7.1. Draw the circuit of a simple medium-frequency *L-C* oscillator using either a transistor or a fet.

   If such an oscillator has a fixed inductance of 60 μH and it is required to tune over the band 1–2 MHz, calculate the range of the variable capacitor to be used.

7.2. Explain how oscillations are maintained in a simple tuned-drain oscillator; where does the oscillatory power come from? An oscillator is to be designed to cover the frequency range 10 kHz to 100 kHz. If the single tuning capacitor used has a maximum-to-minimum capacitance range of 4:1, how many coils will be needed?

7.3. Sketch the circuit and outline the operation of a mutual inductance coupled transistor oscillator. Such an oscillator uses a 50 μH fixed inductor tuned by a variable capacitor. Calculate the range of capacitance necessary to tune the oscillator over the band 3–9 MHz. (C&G)

7.4. Sketch the circuit of a mutual inductance coupled transistor oscillator. Briefly explain the principle of operation. The tuned circuit of such an oscillator is resonant at 1.43 MHz. If the value of the inductance is increased by 69% what is the new frequency of resonance? (C&G)

7.5. Sketch the circuit diagram of a mutual inductance coupled transistor oscillator. Briefly explain (*a*) the biasing and stabilization arrangements, (*b*) how the oscillations are maintained. If such an oscillator has a fixed inductance of 100 μH and it is required to tune over the band 2–4 MHz, calculate the range of the variable capacitor used. (C&G)

7.6. Draw the circuit of a tuned-base oscillator. Outline its method of operation, making particular reference to the frequency-determining components, the bias arrangements, the power supply, and the point from which the output may be obtained.

7.7. (*a*) State whether the transistor is basically a current- or a voltage-operated device. (*b*) Sketch the circuit of a self-biased transistor oscillator incorporating a tuned circuit with feedback by mutual inductance. (*c*) Briefly explain why such an oscillator is self-starting and produces oscillations of constant amplitude. (*d*) If the capacitance of the tuned circuit is halved determine the percentage change in the frequency of oscillation. (C&G)

7.8. An oscillator is required to tune over the frequency range 500–2000 kHz and uses a coil of self-inductance 150 μH. Calculate the maximum and minimum values of the variable tuning capacitor required. (C&G)

7.9. (*a*) What is meant by a *waveform generator*? Draw some waveforms which are in common use and give examples of their application.

   (*b*) A rectangular waveform varying between 0 V and +10 V has the following parameters: p.r.f. 100 kHz, duration of positive pulse 4 μS. What are (i) the mark/space ratio, and (ii) the duty factor of this waveform?

7.10. Draw the block diagram of an oscillator and explain the function of each block. Explain why (i) the loop phase shift must be zero, and (ii) the loop gain must be unity.

Draw the output waveform you would expect if, due to a faulty transistor, the loop gain should fall below unity.

7.11. (i) Draw the circuit diagram of a transistor oscillator designed to work at a fixed frequency. (ii) Indicate component values and fully describe the operation of your circuit. (iii) State the approximate frequency of operation.

7.12. (*a*) Draw the circuit diagram of a tuned-collector transistor oscillator. (*b*) Outline the operation of such an oscillator. (*c*) The tuned circuit of an oscillator consists of a 160 mH fixed inductor in parallel with a variable capacitor. Determine the frequency of resonance when the tuning capacitor is (i) 2500 pF (ii) 625 pF. (C&G)

Short Exercises

7.13. Sketch (i) a sinusoidal waveform of peak value 10 V and frequency 10 kHz, (ii) a square waveform of peak value 10 V and periodic time 100 μs, and (iii) a sawtooth waveform of peak value 10 V and periodic time 100 μs.

7.14. List some common uses for (i) sinusoidal waveforms, (ii) rectangular waveforms, and (iii) sawtooth waveforms.

7.15. Draw the circuit of a tuned-drain fet oscillator. If its drain tuned circuit has an inductance of $10/2\pi$ mH and a capacitance of $100/2\pi$ pF determine its frequency of oscillation.

7.16. (i) What determines the frequency of an oscillator?

(ii) How is the amplitude of the oscillations prevented from continually increasing?

7.17. Explain the meaning of each of the following terms: (i) damped oscillation, (ii) undamped oscillation, and (iii) positive feedback?

7.18. With reference to a tuned-collector oscillator using a p-n-p transistor explain how a leaky-base bias circuit operates to provide Class C bias.

7.19. Draw the circuit diagrams of (i) a series-fed tuned-drain oscillator using a mosfet and (ii) a series-fed tuned-base oscillator using an n-p-n transistor.

7.20. (i) What is meant by the *frequency stability* of an oscillator?

(ii) List the factors which may affect the frequency stability of an oscillator.

7.21. Draw a square wave having a pulse repetition frequency of 1 MHz. What are the mark/space ratio and (ii) the duty cycle of this waveform?

7.22. Answer the following questions on Fig. 7.17. (*a*) What type of transistor is used? (*b*) What polarity should the collector supply voltage be? (*c*) What are the functions of (i) R_1, R_2 and R_4 (ii) R_2, C_1, and (iii) C_3? How can the output be taken from the circuit?

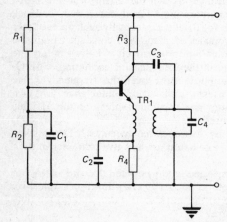

Fig. 7.17

8 Digital Elements and Circuits

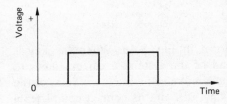

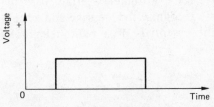

Fig. 8.1 Digital waveforms

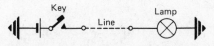

Fig. 8.2 Simple Morse code circuit

Introduction to Codes

Most electronic and telecommunication equipment used for the processing and transmission of information is *analogue* in its nature. That is, the equipment is expected to operate with signals whose amplitude and frequency vary continuously with time. Examples of analogue systems and equipment are radio and television broadcasting and the corresponding receivers in the home, and the telephone system with the telephone receiver. However, electronics can be employed for many other purposes than telecommunication and very often these other uses are *digital* in nature. This means that the signals to be processed are not continuous but, instead, are in the form of pulse trains with varying characteristics. Examples of such digital equipments are digital computers, digital voltmeters, digital watches, pocket calculators, and various control circuitry, e.g. traffic light and lift controls. Also, increasingly nowadays analogue information signals, such as speech and music waveforms, are converted into an equivalent digital form before being transmitted over a radio or a line link.

In a digital equipment or a digital transmission system, information is transmitted in the form of a number of PULSES. The pulses may have one of two values: either a positive or a negative voltage, or they may vary between zero volts and either a positive or a negative voltage. Each pulse can only be in one of two states and Fig. 8.1 shows two possible pulse waveforms. For pulse waveforms to carry information some kind of a CODE must be utilized. The two codes which are common in telegraphy are the *Morse code* and the *Murray code*, while for digital equipments the *Binary code* is used.

A simple example of a digital signalling system using the Morse code is shown in Fig. 8.2. An earthed battery and a key are connected to one end of a telephone line and an earthed lamp is connected to the other end. When the key is unoperated the circuit is broken and no current flows into the line; the lamp is not lit. When the key is pressed, the circuit is completed and the current flowing through the lamp causes it to light. The lamp and the switch are examples of *two-state devices*. For information to be transmitted over the line the key must be operated in accordance with the Morse code.

In the MORSE CODE, characters are represented by a combination of dot signals and dash signals; the difference between a dot and a dash is one of time duration only, a dash being three times the length of a dot. Spacings between elementary signals, between letters and between words are also distinguished from one another by different time durations. As an example, Fig. 8.3 shows the word BAT in Morse code—B is dash and three dots, A is dot, dash, and T is just dash.

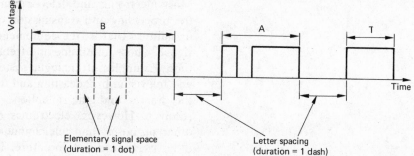

Fig. 8.3 BAT in Morse code

Elementary signal space
(duration = 1 dot)

Letter spacing
(duration = 1 dash)

A better system is shown in Fig. 8.4a. When the key is pressed at the sending end of the system, a current flows to line and passes through the windings of a telegraph relay at the receiving end. The relay operates and its contact completes a circuit for the buzzer to operate. If the key is operated in accordance with the Morse code, a trained operator listening to the buzzer will be able to recognize the message and write it down. Fig. 8.4b shows the essentials of a radio telegraphy

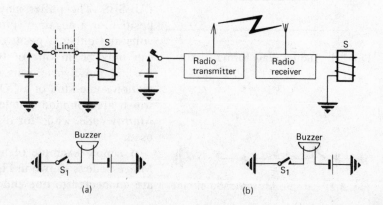

Fig. 8.4 Morse telegraphy systems over (a) a telephone line, (b) a radio link

system. When the sending end key is opened, the radio transmitter is switched off and no signal is radiated; when the key is pressed, the transmitter radiates a single-frequency tone. If the key is operated in accordance with the Morse code, an interrupted continuous wave is radiated and is received by the distant radio receiver. The radio receiver converts the received signal into pulses of direct current that operate the buzzer relay.

The Morse code suffers from the disadvantage that the number of signal elements needed to indicate a character is not the same for all characters and also the signal elements themselves are of different lengths. This makes the design of automatic printing receiving equipment difficult and for teleprinter systems the Murray code is used.

In the MURRAY CODE, all characters have exactly the same number of signal elements and the signal elements are of constant length. Each character is represented by a combination of five signal elements that may be either a mark or a space. In Great Britain a mark is represented by a negative potential or the presence of a tone, and a space is represented by a positive potential or the absence of a tone. It is possible to indicate 2^5, or 32, combinations directly by the Murray code but this number is insufficient for general use, because 26 combinations are required for the letters of the alphabet and there are a number of figures and punctuation marks also to be transmitted. Two combinations, therefore, are used as letter-shift and figure-shift signals and they have the function of setting up the receiving teleprinter to print either figures or letters. This arrangement means that the other combinations can each be used to represent two different characters and so the capacity of the code is greatly increased. Fig. 8.5 shows the word BAT in Murray code; to maintain synchronism between the transmitting and receiving mechanisms, each character is preceded by a start signal (equal to one space) and followed by a stop signal (equal to one and a half marks).

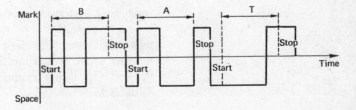

Fig. 8.5 BAT in Murray code

Short-distance data transmission links also use signals generated by a teleprinter but often in conjunction with a different code that is able to accommodate more characters.

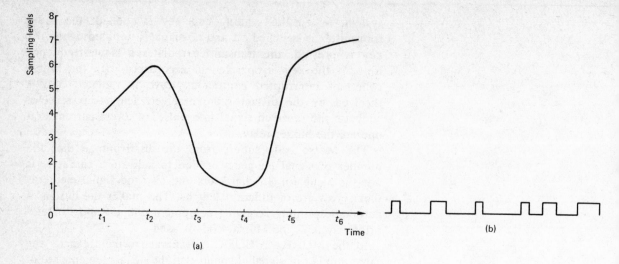

Fig. 8.6 Principle of pulse code modulation

The Morse and Murray codes can only be used for the transmission of written messages. There are cases in practice where it is an advantage to transmit a speech signal over a digital circuit, but in order to do so it will first be necessary to convert the signal from analogue into digital form. This can be achieved using a system known as *pulse code modulation*, the basic principle of which is illustrated by Fig. 8.6. The range of voltages, both positive and negative, over which the analogue signals may vary is divided into a number of *sampling levels*, eight in Fig. 8.6a. The signal amplitude is electronically sampled at each of the time instants marked, i.e. t_1, t_2, t_3, etc., and information about the instantaneous amplitude of each sample is transmitted in digital form using the Binary code.

The Binary Code

In digital electronic systems, the devices have two stable states, ON and OFF, and for this reason the binary number system (or base-2 system) is used. In the binary system only two digits 0 and 1 exist. Larger numbers are obtained by utilizing the powers of two. The digit at the right-hand side of a binary number represents a multiple (0 or 1) of 2^0; the next digit to the left represents a multiple of 2^1; and so on as shown by Table 8.1.

Table 8.1

2^6	2^5	2^4	2^3	2^2	2^1	2^0
64	32	16	8	4	2	1

The value of each power of two is given in the table and any desired number can be attained by a correct choice of zeros and ones. Thus, number 18 for example: 18 is equal to 16 plus 2 and is therefore given by 0010010 in a 7-digit code or by 10010 if only 5 digits are used. Reading from the right (the *least significant digit*), the number consists of zero 1, one 2, zero 4, zero 8 and one 16. Similarly, the binary equivalents of some other numbers are given in Table 8.2, assuming a 5-digit code.

Table 8.2

7	00111	25	11001
17	10001	31	11111

In a pulse code modulation system, the digits 0 and 1 are represented by pulses. In this chapter the binary code will let zero pulse represent binary 0 and a positive pulse represent binary 1, but other codes are also used. Refering to Fig. 8.6a, at time t_1 the amplitude of the sample is level 4, at t_2 it is level 6, and so on. The binary pulses transmitted to line to represent these sampling levels are shown by Fig. 8.6b. In between each train of pulses representing a particular sampling, a zero voltage gap, equal in length to the time duration of a single pulse, has been left. In a practical system this gap is occupied by synchronization information.

In many other fields the use of digital signals is widespread. Some typical examples of digital systems are the following: (*a*) the digital computer, and communication with a computer over telephone lines; (*b*) the control of traffic lights, of lifts in tall buildings, of conveyor belts in factories, etc.; (*c*) safety arrangements for cranes and various factory machines; and (*d*) electronic counting systems such as counting the number of cars in a car park.

Two-state Devices

A two-state device is one which has only two stable operating conditions, or states. Three examples of two-state devices have already been introduced in this chapter, namely the lamp, the switch and buzzer. In each of these cases the device is either operated or it is not, i.e. either it is ON or it is OFF. The two binary conditions 1 and 0 can therefore be represented by a two-state device. Either of two conventions can be adopted; logic-1 can be represented by the ON condition and logic-0 by the OFF state, or alternatively, the ON state can mean logic-0 and the OFF state logic-1.

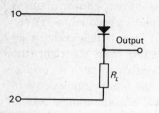

Fig. 8.7 The diode as a two-state device

Table 8.3

Terminal 1	Terminal 2	Output Terminal
Positive	Negative	Positive
Negative	Positive	Positive
Positive	Positive	Positive
Negative	Negative	Negative

The Semiconductor Diode

A semiconductor diode is able to operate as a two-state device because it offers a low resistance to the flow of an electric current in one direction and a high resistance in the other. The diode is said to be ON when it is forward biased and OFF when it is reverse biased. To see how a diode acts as a two-state device consider the circuit of Fig. 8.7 which shows a diode connected in series with a load resistor R_L.

When terminal 1 is positive with respect to terminal 2, the diode conducts and a current flows to develop a voltage across R_L. Neglecting the small voltage drop which must occur across the diode, the voltage appearing at the output terminals of the circuit will be equal to the voltage applied to terminal 1. When the voltage applied to terminal 2 is positive relative to the voltage at terminal 1, the diode will not conduct. The voltage at the output terminal will now be equal to the voltage at terminal 2. Suppose now that voltages of the same magnitude and polarity are applied to terminals 1 and 2. The diode will not conduct and the output voltage will be the same as the common value of the input voltages. The action of the circuit can be expressed by a *voltage table*, as Table 8.3.

The Transistor

Fig. 8.8 shows a typical set of output characteristics for a transistor. A d.c. load line has been drawn on the characteristics between the points $V_{ce} = E_{cc}$, $I_c = 0$ and $I_c = E_{cc}/R_L$, $V_{ce} = 0$. When a transistor is used as an amplifying device, its operation is restricted to part of its characteristics in order to minimize distortion of the applied signal. When used as a switch, a transistor is rapidly switched between two states. When the base current is zero, or perhaps positive (negative for a n-p-n transistor), the transistor is held in its OFF condition. When the transistor is OFF it conducts a very small collector current, equal to the collector leakage current or less. The voltage dropped across the collector resistor is negligible and so the voltage across the transistor in the OFF state is equal to the collector supply voltage. When the transistor is driven into saturation, and is often said to be *bottomed*, it is in its ON state. The voltage across the transistor is now its *saturation voltage* and this is only a fraction of a volt.

The transistor can be switched rapidly between its ON and OFF states by the application of a rectangular waveform of sufficient amplitude to its base terminal (Fig. 8.9). When the input waveform is at zero potential with respect to earth, the transistor will be switched into its OFF state; the voltage which appears at the output terminals is then equal to the collector supply voltage since there will be zero voltage drop across R_L.

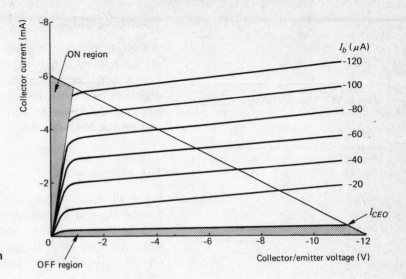

Fig. 8.8 The transistor as a switch

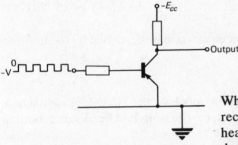

Fig. 8.9 Switching a transistor

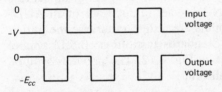

Fig. 8.10 Input and output waveforms of the transistor switch

When the transistor is switched into its ON state by the rectangular base signal voltage, the transistor will conduct heavily and a large voltage, approximately equal to E_{cc}, is dropped across R_L. The output voltage of the circuit is then equal to the saturation voltage $V_{ce(sat)}$ of the transistor; since $V_{ce(sat)}$ is very small it is customary to assume it to be zero. The voltage at the output terminals of the circuit switches between zero volts and $-E_{cc}$ volts. It should be noted that when the input signal voltage is negative the output voltage is zero, and when the input signal voltage is zero the output voltage is $-E_{cc}$ volts (see Fig. 8.10). This means that the input waveform has been inverted; it will be seen later that this means that the circuit has performed the logical function NOT. If an n-p-n switching transistor is considered, a pulse waveform varying between 0 and +V volt must be applied to the base to switch the output voltage between 0 and E_{cc}.

The more positive (less negative) voltage level can be regarded as representing logic-1 and the less positive level as giving logic-0. This convention is known as POSITIVE LOGIC. Conversely, logic-1 can be taken as being the least positive, or more negative, voltage level with the more positive voltage being labelled as logic-0. This second convention is known as NEGATIVE LOGIC. Clearly, the convention employed in a particular case must be clearly stated, or understood.

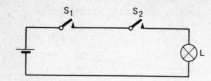

Fig. 8.11 The AND-logic function

Table 8.4

Switch S_1	0	1	0	1
Switch S_2	0	0	1	1
Lamp L	0	0	0	1

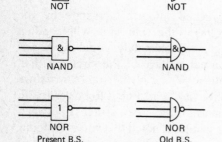

Fig. 8.12 Gate symbols

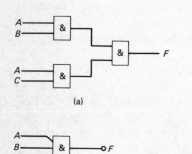

Fig. 8.13

Electronic Gates

An electronic gate is a logic element which is able to operate on an applied binary signal in a manner determined by its logical function. A number of different types of gate feature in digital circuitry, the most common of which are (i) the AND-gate, (ii) the OR-gate, (iii) the NOT-gate, (iv) the NOR-gate, and (v) the NAND-gate.

The AND-gate

Fig. 8.11 shows a lamp connected in series with two switches, S_1 and S_2, and a battery. For a current to flow in the circuit and the lamp to light both S_1 *and* S_2 must be closed. A switch when open, or OFF, is denoted by logic-0 and when closed, or ON, is represented by logic-1, while the state of the lamp when it is lit is given by logic-1 and when unlit by logic-0. The operation of the circuit can then be described by its TRUTH TABLE (see Table 8.4).

The action of the circuit can also be described by the Boolean equation:

$$L = S_1 \cdot S_2$$

The dot $\cdot$ is the Boolean symbol for the AND logical function. Equation (8.1) states that, for the lamp L to be ON or 1, both S_1 *and* S_2 must be ON or 1.

An AND-gate is the electronic equivalent of switches connected in series and is a circuit having two or more input terminals and one output terminal. The number of inputs to a gate is known as its FAN-IN. The output state of an AND-gate is 1 only if *all* the inputs are also 1, otherwise the output will be at 0. The British Standards Institution (B.S.I.) symbol for an AND-gate is given in Fig. 8.12. The previous symbol, now superseded, is also shown.

EXAMPLE 8.1

Fig. 8.13*a* shows a digital circuit constructed using AND-gates. (*a*) Obtain the Boolean equation of the circuit. (*b*) Write down the truth table of the circuit. (*c*) Simplify the circuit.

Solution
(*a*) Let the output of the top left-hand gate be D, then $D = A \cdot B$. The output of the bottom left-hand gate is $E = A \cdot C$. The output of the circuit is $F = D \cdot E = A \cdot B \cdot A \cdot C$

Since $A \cdot A = A$ $F = A \cdot B \cdot C$ (*Ans.*)

(b) The truth table is

A	B	A · B	C	A · C	F
0	0	0	0	0	0
1	0	0	0	0	0
0	1	0	0	0	0
0	0	0	1	0	0
1	1	1	0	0	0
1	0	0	1	1	0
0	1	0	1	0	0
1	1	1	1	1	1

(c) The required logical function could be produced by a single 3-input AND-gate as shown in Fig. 8.13b.

The truth table for a 3-input AND-gate is given by Table 8.5; it is clear that $F = 1$ only if A AND B AND C are each 1.

Table 8.5

A	0	1	0	0	1	1	0	1
B	0	0	1	0	1	0	1	1
C	0	0	0	1	0	1	1	1
F	0	0	0	0	0	0	0	1

The OR-gate

Current will flow in the circuit shown in Fig. 8.14 and light the lamp if either switch S_1 OR switch S_2 OR both S_1 and S_2 are closed or ON. Table 8.6 is the truth table for this circuit which acts as a two-input OR-gate.

Table 8.6

Switch S_1	0	1	0	1
Switch S_2	0	0	1	1
Lamp	0	1	1	1

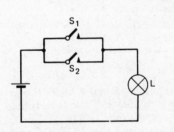

Fig. 8.14 The OR-logic function

The Boolean equation for the circuit is

$$L = S_1 + S_2 \qquad (8.2)$$

The + symbol represents the logical function OR and the circuit symbol for an OR-gate is given in Fig. 8.12.

EXAMPLE 8.2

The rectangular waveforms shown in Fig. 8.15 are applied to the inputs of (a) a two-input AND-gate and (b) a two-input OR-gate. Draw the output waveform of each gate. Assume positive logic.

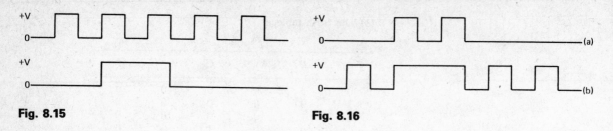

Fig. 8.15

Fig. 8.16

Solution

(a) The output of the AND-gate will be 1 only when both of its input waveforms are 1. Fig. 8.16a shows the output waveform.

(b) The output of the OR-gate will be 1 when either or both of its input waveforms are 1. Hence the output waveforms will be as given by Fig. 8.16b.

EXAMPLE 8.3

Determine the Boolean equations that describe the logic circuits whose truth tables are given in Tables 8.7a and b. Draw the circuits.

Table 8.7

A	B	C	F	A	B	C	F
	(a)				(b)		
0	0	0	0	0	0	0	0
1	0	0	1	1	0	0	0
0	1	0	1	0	1	0	0
0	0	1	1	1	1	0	1
1	1	0	1	0	0	1	1
1	0	1	1	1	0	1	1
0	1	1	1	0	1	1	1
1	1	1	1	1	1	1	1

Solution

(a) The output F is 0 only if all three inputs A, B and C are 0. If one or more of the inputs is 1, the output is 1. Hence, Table 8.7a is the truth table of a three-input OR-gate (Fig. 8.17a).

(b) The output F is 1 either if A *and* B are 1 *or* if C is 1 OR if A *and* B and C are 1. This means that inputs A and B are connected to a two-input AND-gate whose output is applied, with input C, to a two-input OR-gate (see Fig. 8.17b).

Fig. 8.17

EXAMPLE 8.4

(a) Write down the Boolean equations that describe the logic circuits given in Figs. 8.18a and b. (b) Write the truth table for each circuit and compare them. Comment on the results.

Solution

(a) For circuit a, $F = A \cdot B + B \cdot C$ and for circuit b, $F = (A + C) \cdot B$

(b) The truth table for circuit a is

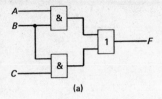

(a)

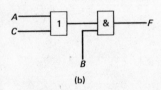

(b)

Fig. 8.18

A	0	1	0	0	1	1	0	1
B	0	0	1	0	1	0	1	1
$A \cdot B$	0	0	0	0	1	0	0	1
C	0	0	0	1	0	1	1	1
$B \cdot C$	0	0	0	0	0	0	1	1
F	0	0	0	0	1	0	1	1

and the truth table for circuit *b* is

A	0	1	0	0	1	1	0	1
B	0	0	1	0	1	0	1	1
C	0	0	0	1	0	1	1	1
$A + C$	0	1	0	1	1	1	1	1
F	0	0	0	0	1	0	1	1

Comparing the rows in the tables that give the outputs *F* of the two circuits it can be seen that they are identical. This means that the same logical function can often be produced by different combinations of gates. In this example, circuit *b* requires only two gates to perform the desired function as opposed to the three gates needed by circuit *a*, and so circuit *b* would be chosen.

The NOT-gate

The NOT logical function is performed by the circuit in Fig. 8.19. Before the switch S_1 is operated, current flows in the circuit and the lamp is lit, or ON. Operation of the switch stops current flowing in the circuit and turns the lamp OFF. The truth table for this circuit is given by Table 8.8.

Table 8.8

S	0	1
L	1	0

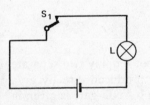

Fig. 8.19 The NOT function

The Boolean equation of the circuit is

$$L = \bar{S} \tag{8.3}$$

The bar over a symbol means "NOT that symbol." In the equation, it means "NOT S." The symbol for a NOT-gate is given in Fig. 8.12.

EXAMPLE 8.5

(a) Write down the truth table of the circuit of Fig. 8.20.
(b) Write down the Boolean equation representing the circuit.

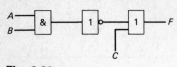

Fig. 8.20

Solution

(*a*) The truth table of the circuit is

A	0	1	0	1	0	1	0	1
B	0	0	1	1	0	0	1	1
$A \cdot B$	0	0	0	1	0	0	0	1
$\overline{A \cdot B}$	1	1	1	0	1	1	1	0
C	0	0	0	0	1	1	1	1
F	1	1	1	0	1	1	1	1

(*b*) $F = \overline{A \cdot B} + C$ (*Ans.*)

The NOT logic function is produced by the transistor circuit shown in Fig. 8.9.

The NAND-gate

Fig. 8.21 shows a circuit that consists of an AND-gate followed by a NOT-gate; the truth table for this circuit is given by Table 8.9.

Table 8.9

A	0	1	0	0	1	1	0	1
B	0	0	1	0	1	0	1	1
C	0	0	0	1	0	1	1	1
$A \cdot B \cdot C$	0	0	0	0	0	0	0	1
$\overline{A \cdot B \cdot C}$	1	1	1	1	1	1	1	0

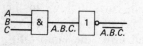

Fig. 8.21   The NAND-gate

The NOT AND logic function need not employ two separate gates as shown, but, instead, can be produced directly by a NAND-gate. The circuit symbol for a NAND-gate is given in Fig. 8.12.

The NOR-gate

The NOT OR or NOR-gate performs the same logical operation as an OR-gate followed by a NOT-gate (Fig. 8.22). The circuit symbol for a NOR-gate is given in Fig. 8.12.

The NOT logic function can be produced by either a NAND-gate or a NOR-gate, the former gate requiring all its input terminals to be connected to the input signal line.

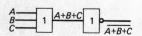

Fig. 8.22   The NOR-gate

Diode Logic Gates

The circuit diagram of a three-input diode logic gate is shown in Fig. 8.23. The positive voltage E applied to the resistor R is normally several times greater than the peak positive voltage V of the pulse waveforms which are applied to input terminals A, B and C. If any one of the inputs is at zero potential, the

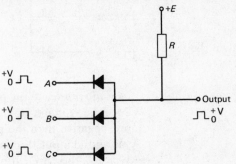

Fig. 8.23 The diode-resistor gate

associated diode conducts heavily until the voltage dropped across resistor R reduces the output voltage to zero. The other two diodes now have positive voltage V at their cathode and zero voltage at their anode, and so they do not conduct. Similarly, if two inputs are at zero potential, the diode associated with the third input will be reverse-biased and the output voltage will be zero. When all three inputs are simultaneously at zero potential, all three diodes will conduct and again the output voltage will be zero. Now suppose that the three inputs have a positive pulse applied to them at the same time. The three associated diodes will conduct, since each one has a more positive voltage at its anode. Current will flow in resistor R and the output voltage will fall until it is equal to the voltage of the input pulses, i.e. V volts.

The voltage table of the diode logic gate is given by Table 8.10.

Table 8.10

A	0	+V	0	0	+V	+V	0	+V
B	0	0	+V	0	+V	0	+V	+V
C	0	0	0	+V	0	+V	+V	+V
O/P	0	0	0	0	0	0	0	+V

The logical function of the circuit depends upon whether positive or negative logic is employed. With positive logic the circuit acts as an AND-gate, since only when all three inputs are at positive voltage V is the output at voltage V. Conversely, if negative logic is used the circuit provides the OR function.

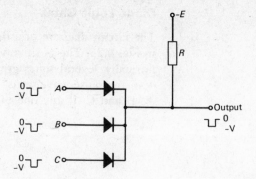

Fig. 8.24 An alternative diode-resistor gate

An *alternative form* of diode logic circuit is shown in Fig. 8.24. The voltage applied to the resistor R is considerably more negative than the peak value of the pulses applied to the A, B and C input terminals. When all three input pulses are at $-V$ volts, the three diodes conduct until the output voltage becomes $-V$ volts. If one or more of the input pulses is at zero potential, the associated diode conducts until the output voltage has fallen to zero. The other diodes then have a negative voltage at their anodes and zero voltage at their cathodes and so are turned OFF. When all three input pulses are simultaneously at zero volts, the three diodes are switched ON and the output voltage is clamped to zero potential. Table 8.11 shows the voltage table for the circuit. It can be seen that the circuit acts as an OR-gate with positive logic and as an AND-gate with negative logic.

Table 8.11

A	0	$-V$	0	0	$-V$	$-V$	0	$-V$
B	0	0	$-V$	0	$-V$	0	$-V$	$-V$
C	0	0	0	$-V$	0	$-V$	$-V$	$-V$
O/P	0	0	0	0	0	0	0	$-V$

DIODE-RESISTOR LOGIC gates cannot perform the NOT logical function and hence are also unable to provide either the NAND or the NOR functions. To obtain either of these functions the diode gate must be followed by an inverting transistor stage as shown in Fig. 8.25.

Whenever the voltage applied to the base of the transistor is at $+V$ volts (see Table 8.10), the transistor will be driven into saturation and its collector voltage will be approximately zero. Conversely, when the base voltage is at $0 \, V$ the transistor will be turned OFF and the output voltage will be $+E_2$ volts. The voltage table for the circuit is given by Table 8.12. A comparison of Table 8.12 with Tables 8.10 and 8.11 will show that

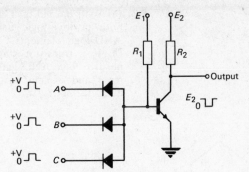

Fig. 8.25 Use of a transistor inverting stage

when positive logic is used, the circuit performs the NAND function, and when negative logic is used it is a NOR-gate.

Table 8.12

A	B	C	X	F
0	0	0	0	$+E_2$
$+V$	0	0	0	$+E_2$
0	$+V$	0	0	$+E_2$
0	0	$+V$	0	$+E_2$
$+V$	$+V$	0	0	$+E_2$
0	$+V$	$+V$	0	$+E_2$
$+V$	0	$+V$	0	$+E_2$
$+V$	$+V$	$+V$	$+V$	0

In practice, diode logic circuitry is rarely used since it suffers from a number of disadvantages. These are as follows:

(a) The forward resistance of a diode is not zero and so there is a voltage drop across each conducting diode. This has the effect of reducing the high voltage level and increasing the low voltage level; this reduces the number of gates which can be cascaded to perform a required logical function.

(b) The maximum speed of switching between logic states is low compared with alternative gate circuits.

(c) The NOT function cannot be performed.

(d) The load resistance R is shunted by the impedance of circuits that are connected to the output terminals. This will have the effect of altering the output voltage levels.

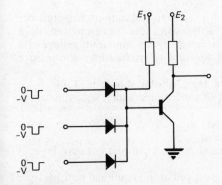

Fig. 8.26

Exercises

8.1. With the aid of a circuit diagram explain the operation of a 2-input diode-transistor NAND-gate using a p-n-p transistor.

8.2. Fig. 8.26 shows the circuit of a logic gate.

(a) Describe the operation of the circuit.

(b) State whether the transistor should be a p-n-p or a n-p-n type and determine the polarities of the voltages E_1 and E_2.

(c) State the logical function of the circuit if positive logic is employed.

8.3. (*a*) With the aid of a circuit diagram, explain the operation of both positive-logic and negative-logic semiconductor diode AND-logic elements.

(*b*) If a transistor stage is connected to the output of the positive-logic AND-gate drawn for part (*a*), sketch the arrangement and explain its operation. What logic function is now being formed? (C&G)

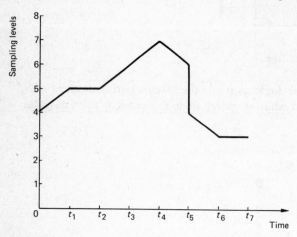

Fig. 8.27

Fig. 8.28

8.4. The waveform shown in Fig. 8.27 is to be transmitted using pulse code modulation using 8 sampling levels. Draw the pulse waveform that is transmitted to line. Assume each pulse train to be separated by an interval equal to the length of two binary pulses.

8.5. (*a*) Draw circuits, using AND, OR and NOT gates only, to carry out the following logical functions:

(i) $F = \overline{A + B + C \cdot D}$

(ii) $F = (A + B) \cdot (C + D) \cdot (\bar{A} + B)$

(*b*) Fig. 8.28 shows a logic circuit. Determine whether A should be 1 or 0 to produce

$F = 1$ when (i) $C = 1$ and $B = 0$
(ii) $C = 0$ and $B = 1$

8.6. Describe, with the aid of a set of typical output characteristics and a load line, how a transistor can be employed as a two-state device. What is meant by the *saturation voltage* of a transistor? What logical function is produced by a transistor switch?

8.7. The voltage table for a logic gate is shown in Table 8.13. Determine, with the aid of truth tables, which logic function is produced if

(i) positive logic is used,
(ii) negative logic is used,
(iii) positive logic is used at the input and negative logic at the output, and
(iv) negative logic is employed at the input and positive logic at the output.

Table 8.13

A	B	F
−6 V	−6 V	−1 V
−1 V	−6 V	−1 V
−6 V	−1 V	−1 V
−1 V	−1 V	−6 V

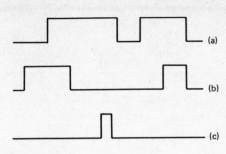

Fig. 8.29

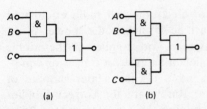

(a)   (b)

Fig. 8.30

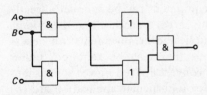

Fig. 8.31

8.8. Draw and explain the operation of a semiconductor diode AND-logic element for use with positive logic. What will the circuit be if negative logic is used?

8.9. The waveforms shown in Fig. 8.29 are applied to the input of
(i) a three-input AND-gate, and
(ii) a three-input OR-gate.
For each case determine the output waveform.

Short Exercises

8.10. Write down the truth table for (i) a three-input OR-gate, (ii) a three-input AND-gate, and (iii) a NOT-gate. Also give the corresponding Boolean expressions.

8.11. (*a*) What is meant by the binary representation of a number? (*b*) Convert the following decimal numbers into binary: 37, 49, 71 and 130.

8.12. What is meant by a *two-state device*? Give six examples.

8.13. Draw circuits to perform the logical operations
(i) $F = A \cdot B \cdot C + D \cdot E$
(ii) $F = (A + B + C) \cdot (D + E)$

8.14. Explain, with the aid of suitable diagrams, how information can be transmitted over a telephone line using a pulse waveform.

8.15. The readings of a voltmeter at an unattended electrical installation are to be transmitted over a telephone line to a distant staffed point. Suggest a way in which this might be achieved. Assume that the readings will always be in the range 0–10 V and that the readings rounded off to the nearest volt are sufficiently accurate.

8.16. What is meant by (i) positive logic and (ii) negative logic? Illustrate your answer with an example.

8.17. Write down the truth table and the Boolean equation of the circuits shown in Figs. 8.30 *a* and *b*.

8.18. Repeat Example 8.2 (page 117) using negative logic.

8.19. Determine whether input A should be at logic 1 or 0 to produce an output 1 from the circuit shown in Fig. 8.31.

9 Power Supplies

For the correct operation of electronic and radio equipment suitable power supplies must be available; d.c. power is required for the anode and screen grid supplies of thermionic valve equipment and the collector supplies of transistor equipment; and a.c. power is required for the heater supplies of thermionic valve equipment. Rarely are the correct supplies available and power supply equipment is necessary to convert an available supply into the wanted supply. The d.c. requirements of a small transistor equipment can be supplied directly by a small dry battery, but for larger transistor and thermionic valve equipments more complicated arrangements are necessary. The power conversions that may be required, either singly or in combination, to convert a given power source into the power supply required for a particular application are (a) the conversion of a.c. from one voltage to another, (b) the conversion of a.c. into d.c., (c) the conversion of d.c. into a.c., and (d) the conversion of d.c. from one voltage to another. The first of these conversions can be easily achieved by the use of a transformer of the appropriate turns ratio; methods of achieving the remaining conversions form the subject of this chapter. Only relatively low-power supplies, such as those required for radio receivers, tape recorders and other small equipments, are to be discussed.

Rectifier Circuits

A number of different circuits exist that are capable of converting an a.c. supply into a pulsating d.c. current and they may be broadly divided into one of two classes, half-wave rectifiers and full-wave rectifiers. In the circuits that follow, diodes (thermionic valve or semiconductor) may be used as the rectifying device, but of course the valve diodes require a heater supply.

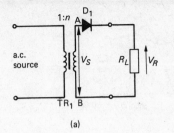

(a)

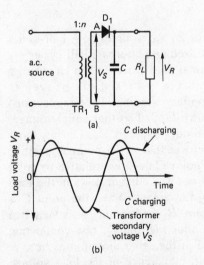

(b)

Fig. 9.1 The half-wave rectifier with resistance load

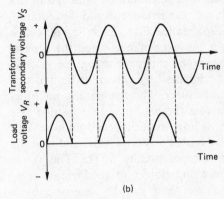

(a)

(b)

Fig. 9.2 The half-wave rectifier with resistance-capacitance load

Half-wave Rectification

In its simplest form, half-wave rectification consists merely in the connection of a diode in series with the a.c. supply and the load, as shown in Fig. 9.1a.

The diode conducts only during those alternate half-cycles of the a.c. supply voltage V_s that make point A positive relative to point B, and so the load current consists of a series of half sine-wave pulses. The voltage V_R developed across the load is the product of the load current and the load resistance and has the same waveform as the load current (Fig. 9.1b). The disadvantage of this simple rectifying circuit is very clear: the load voltage, although unidirectional, varies considerably and is, indeed, zero for half the time. Such a waveform is only suitable for simple applications, such as battery charging, since the variations will appear as noise at the output of any equipment fed by the supply. When the diode is non-conducting, the peak voltage across it, known as the *peak inverse voltage* (p.i.v.), is equal to the peak value of the transformer secondary voltage. This voltage must not exceed the voltage rating of the type of diode employed.

The d.c. output of a rectifier circuit is required to be as steady as possible and a great step towards this goal could be achieved if the load voltage could be prevented from falling to zero during alternate half-cycles. One way of achieving this is to connect a capacitor C in parallel with the load as shown in Fig. 9.2a. Each time the diode conducts, the current that flows charges the capacitor and the voltage across the capacitor builds up. During the intervals of time when the diode is non-conducting, the capacitor discharges via the load resistance and prevents the load voltage falling to zero (Fig. 9.2b). The capacitor continues to discharge, at a rate determined by the time constant CR_L seconds, until the point A is taken more positive than the capacitor voltage by a positive half-cycle of the input voltage V_s. The diode then conducts, the capacitor is recharged, and the capacitor voltage rises again. If the load current is fairly small the capacitor does not discharge very much between charging pulses and the average load voltage V_R is only slightly less than the peak value of the applied voltage V_s. The p.i.v. is increased to twice the peak secondary voltage V_s because of the polarity of the voltage developed across C.

The value of the load voltage is adjustable, within limits, by suitable choice of the turns ratio n of the input transformer. An increase in the load, i.e. in the load current, means that the load resistance is less; this, in turn, means that the time constant of the discharge path is smaller. Capacitor C then discharges more rapidly and the load voltage is not as constant (see Fig. 9.3).

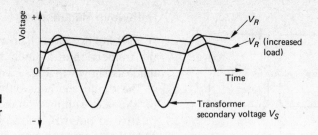

Fig. 9.3 Showing the effect of load changes on a half-wave rectifier with resistance-capacitance load

A completely steady load voltage cannot be obtained in this way since too large a value of capacitance would be required. The maximum value of capacitance that can be employed is limited, because the larger the capacitance value the greater the current required to charge the capacitor to a given voltage, and the current that can be handled by a diode is limited to a figure quoted by the manufacturer. The fluctuating, unidirectional voltage appearing across the load may be regarded as a d.c. voltage having an a.c. voltage superimposed upon it. This a.c. voltage is known as the RIPPLE VOLTAGE and is at the frequency of the supply voltage, usually 50 Hz. The ripple voltage is undesirable, since the object of rectification is to provide a steady d.c. voltage, and can be removed by a smoothing or filter circuit connected between the diode and the load.

Full-wave Rectification

With full-wave rectification of an a.c. source, both half-cycles of the input waveform are utilized and alternate half-cycles are inverted to give a unidirectional load current. The circuit of a full-wave rectifier is shown in Fig. 9.4a and can be seen to require two diodes. The secondary winding of the input transformer TR_1 is accurately centre-tapped so that equal voltages are applied across the two diodes D_1 and D_2. During those half-cycles of the input waveform that make point A positive with respect to point B and point C negative relative to point B, D_1 conducts and D_2 does not and current flows in the load in the direction indicated by the arrow. When the point C is positive with respect to point B, and point A is negative relative to point B, D_2 is conducting and D_1 non-conducting and current flows in the load in the same direction as before. The waveform of the current, and hence of the load voltage V_R, is shown in Fig. 9.4b. The p.i.v. is $2V_s$.

A more constant value of load voltage can be obtained by the connection of a capacitor across the load as shown in Fig. 9.5a. The action of the reservoir capacitor is exactly the same as in the half-wave circuit but now the capacitor is re-charged twice per input cycle instead of only once. Between charging

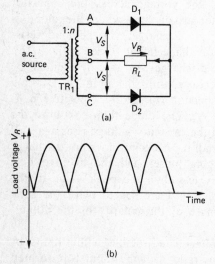

Fig. 9.4 The full-wave rectifier with resistance load

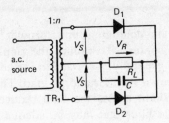

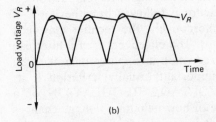

Fig. 9.5 The full-wave rectifier with resistance-capacitance load

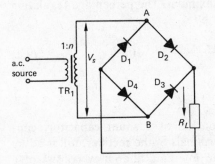

Fig. 9.6 The bridge rectifier

pulses the capacitor starts to discharge through the load but, provided the time constant is not too short, the load voltage has not fallen by much before the next charging pulse occurs (Fig. 9.5b) The load voltage attains a mean value only slightly less than the peak voltage appearing across one half of the input transformer secondary winding. As before, the ripple content of the load voltage increases with increase in load current and can be reduced by the use of a suitable filter network.

The full-wave circuit has a number of advantages over the half-wave circuit: it is more efficient; little, if any, d.c. magnetization of the transformer core occurs; and the ripple voltage is at twice the supply frequency, i.e. at 100 Hz. The increase in the ripple frequency makes it easier to reduce the percentage ripple to a desired level. The disadvantages of the circuit are the need for a centre-tapped transformer and for two diodes.

An *alternative method* of full-wave rectification is the use of a bridge network (see Fig. 9.6). The BRIDGE RECTIFIER circuit requires four diodes instead of two but avoids the need for a centre-tapped input transformer. Further, the arrangement gives a load voltage that is nearly twice as great as that from the circuit of Fig. 9.5a—assuming, of course, the same transformer secondary voltage. During those half-cycles of the input that make point A positive with respect to point B, diodes D_2 and D_4 are conducting and diodes D_1 and D_3 are non-conducting; current therefore flows from point A to point B via D_2, the load R_L, and D_4. When point A is negative relative to point B, D_1 and D_3 are conducting and D_2 and D_4 are non-conducting; current then flows from point B to point A via D_3, the load R_L, and D_1. Both currents pass through the load in the same direction and so a fluctuating, unidirectional voltage is developed across the load having the waveform of Fig. 9.4b. The variations in load voltage can be reduced by the connection of a capacitor across the load; the load voltage waveform is then that of Fig. 9.5b.

When higher d.c. voltages are required, the bridge circuit has some advantages over the circuit using a centre-tapped transformer: the p.i.v. of each diode is only equal to the peak secondary voltage V_s; a centre-tapped secondary winding is not required; and the current-rating of the transformer is less. This means that a smaller and hence cheaper transformer can be used.

Filter Circuits

A power supply unit for electronic equipment must provide a d.c. voltage of minimum ripple content. To reduce the ripple voltage to a tolerable level it is generally necessary to include some kind of filter circuit between a rectifier and its load. The simple capacitor connected in shunt across the load reduces ripple and is, therefore, a simple filter that may be adequate for some applications. Further smoothing of the output voltage can be achieved if the capacitor is followed by either an L-C or an R-C network, or, alternatively, the rectifier may feed directly into a series inductor. The effectiveness of a filter can be judged in terms of the reduction in ripple voltage it gives, but another and equally important (usually) criterion is its voltage regulation. The VOLTAGE REGULATION of a rectifier circuit is a measure of how its output voltage changes as the current taken from it is varied. Ideally, of course, the output voltage should remain constant, so good regulation implies that the voltage changes very little as the load current is varied from minimum to maximum. The percentage regulation of a power supply is given by

$$\% \text{ regulation} = \frac{V_{no\text{-}load} - V_{full\text{-}load}}{V_{no\text{-}load}} \times 100\% \qquad (9.1)$$

Capacitor-input Filters

A capacitor-input filter consists of a shunt capacitor, connected across the output terminals of the rectifier, followed by a basic low-pass filter. The low-pass filter may consist of a series inductor and a shunt capacitor as in Fig. 9.7, or a series resistor and a shunt capacitor as shown in Fig. 9.8. In the L-C filter the value of capacitance C_1 is chosen to give a reasonably smooth output voltage from the rectifier proper, and the values of L and C_2 are chosen to give adequate ripple suppression.

Typical values for C_1 and C_2 for a half-wave rectifier are $32\,\mu\text{F}$ each and for L, $30\,\text{H}$, and at $50\,\text{Hz}$, which is the half-wave rectifier ripple frequency, these components have reactances of $1/(2\pi \times 50 \times 32 \times 10^{-6})$, or approximately $100\,\Omega$, and $2\pi \times 50 \times 30$, or $9426\,\Omega$, respectively. The reactances of C_2 and L act as a potential divider across the rectifier output and reduce the ripple voltage to approximately $100/9326$ times its original value.

In the full-wave rectifier the ripple frequency is $100\,\text{Hz}$ and this means that a filter using the same component values would be approximately four times as efficient. For a given amount of ripple, smaller components can be used and typical values are $C_1 = C_2 = 8\,\mu\text{F}$ and $L = 15\,\text{H}$. Both filters have negligible

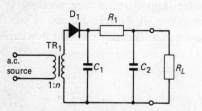

Fig. 9.7 The capacitor-input L-C filter

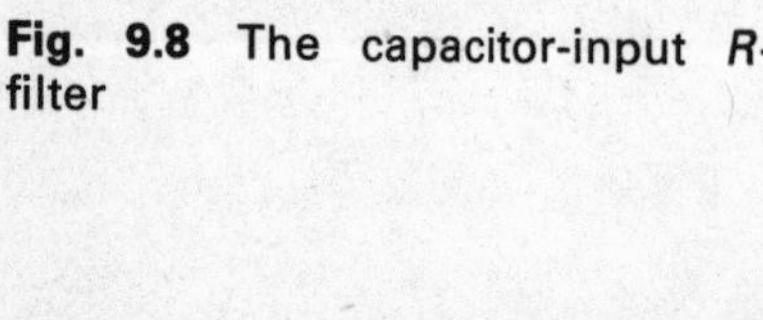

Fig. 9.8 The capacitor-input R-C filter

effect on the wanted d.c. component of the rectified output voltage. In order to obtain the fairly large capacitance values required cheaply and in the minimum of space, electrolytic capacitors are normally used.

The disadvantages of the capacitor-input L-C filter are (a) the cost, weight, size and external fields of the series inductor, and (b) the relatively poor voltage regulation. The first of these disadvantages can be overcome by replacing the series inductor with a series resistor, although this has the obvious disadvantage of increasing the d.c. voltage drop in the filter. The R-C filter, therefore, has poor regulation and requires adequate ventilation to conduct away the heat produced in the resistor. As a result it is only used to supply equipments taking only a small current.

The circuit of a capacitor-input R-C filter is shown in Fig. 9.8, such filters are used in the power supply units of some television receivers and cathode-ray oscilloscopes. Typical values are $R = 100$ to $200\,\Omega$, $C_1 = 100\,\mu F$ and $C_2 = 150$ to $200\,\mu F$.

Choke-input Filters

When a choke-input filter is used there is no reservoir capacitor and the rectifier feeds directly into the filter (Fig. 9.9). Inductor L and capacitor C form a potential divider across the output of the rectifier and reduce the ripple voltage to a low value. The choke-input filter can only be used in conjunction with a full-wave rectifier since it requires current to flow at all times, the current being provided first by D_1 and then by D_2. The current waveforms in the circuit are shown in Fig. 9.10. The fact that current flows continuously, instead of in a series of pulses as in the capacitor-input filter circuits, means that the input transformer is used more efficiently. A further advantage is that the ripple content at the output of the filter is less dependent on the load current. Typical values for L are 5 to 30 H and for C, 5 to 40 μF.

Fig. 9.11 gives a comparison of the voltage regulation of the two types of filter. It can be seen that although the output voltage of the choke-input filter is the smaller, its regulation is better. The regulation can be improved by the use of a larger value of inductance as shown by the two lower curves. If the load current falls below a certain critical value I' or I'' (this could happen each time the equipment was first switched on), the output voltage will rise abruptly. To prevent this happening a resistor, known as a "bleeder," can be connected across the output terminals of the filter to ensure that a current greater than the critical value is always taken.

As an alternative to the use of a bleeder resistor with its

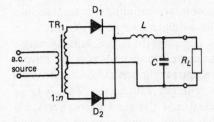

Fig. 9.9 The choke-input filter

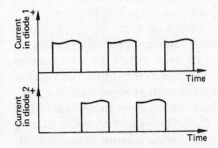

Fig. 9.10 Current waveforms in the choke-input filter

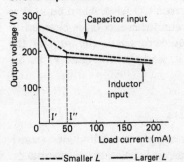

- - - - Smaller L ——— Larger L

Fig. 9.11 Regulation curves for capacitor- and choke-input filters

consequent power dissipation, a *swinging choke* can be used. This is an inductor whose inductance depends upon the magnitude of the direct current flowing in its windings. Typically, such an inductor might have an inductance of 30 H with zero current flow and only 5 H with 250 mA flowing.

Voltage Multiplying

The use of a suitable combination of rectifiers and capacitors can give a d.c. output voltage that is several times greater than the peak voltage appearing across the secondary winding of the input transformer. Consider, for example, Fig. 9.12 which shows a voltage-doubling circuit. During the half-cycles of the input when point A is positive with respect to point B, diode D_1 conducts and capacitor C_1 is charged to the peak voltage V_s appearing across the secondary winding of transformer TR_1. When point B is positive relative to point A, diode D_2 conducts and capacitor C_2 is charged to the same voltage. Capacitors C_1 and C_2 are connected in series across the output terminals and so the voltage appearing across these terminals is equal to twice the peak secondary voltage, i.e. $2V_s$. If large values of capacitance are used and the load current is fairly small, the output voltage has small ripple content and good regulation. Since the capacitors are charged during alternate half-cycles, the ripple frequency is at twice the supply frequency, that is 100 Hz for 50 Hz mains.

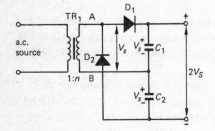

Fig. 9.12 The voltage doubler

The principle of the voltage doubler can be extended to voltage tripling, quadrupling or even higher. Fig. 9.13 shows possible arrangements for (a) a voltage tripler and (b) a voltage quadrupler. Consider the voltage tripler. During half-cycles when point A is positive with respect to point B, diode D_1 conducts and capacitor C_2 is charged to V_s volts. During the half-cycles when point B is positive relative to point A, D_2 conducts and C_1 is charged to $2V_s$ volts—because the voltage applied across it is the sum of the transformer secondary voltage V_s and the voltage V_s across C_2. Also, when point A is positive relative to point B, D_3 conducts and C_3 is charged to $3V_s$, because the voltage applied across it is the sum of transformer secondary voltage V_s and the voltage $2V_s$ across C_1.

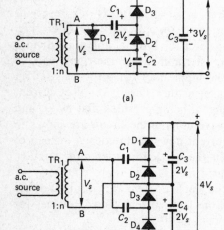

(a)

(b)

Fig. 9.13 (a) The voltage tripler and (b) the voltage quadrupler

Zener Diode Voltage Stabilizers

For many applications a power supply consisting of a transformer, a rectifier and a filter has an inadequate performance. Firstly, the voltage regulation is not good enough and, secondly, the d.c. output voltage varies with change in the a.c.

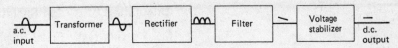

Fig. 9.14 Block diagram of a stabilized power supply

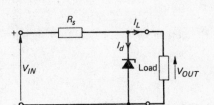

Fig. 9.15 The zener diode stabilizer

input voltage. To improve the constancy of the d.c. output voltage as the load and/or the a.c. input voltage vary, a voltage stabilizer circuit must be employed. The voltage stabilizer is connected between the output of the filter circuit and the load as shown in Fig. 9.14.

A number of transistor stabilization circuits are available but the simplest stabilization circuit consists merely of a resistor R_s connected in series with the input voltage and a zener diode connected in parallel with the load (Fig. 9.15).

Applying Kirchhoff's voltage law to the circuit,

$$V_{IN} = (I_d + I_L)R_s + V_{OUT}$$

Rearranging

$$R_s = \frac{V_{IN} - V_{OUT}}{I_d + I_L} \qquad (9.2)$$

When a zener diode is operated in its breakdown region, the current flowing through the diode can vary considerably with very little change in the voltage across the diode. If the load current should increase, the current through the zener diode will fall by the same percentage in order to maintain a constant voltage drop across R_s and hence a constant output voltage. Should the load current decrease, the diode will pass an extra current such that the sum of the two currents flowing in R_s is maintained constant, and the output voltage of the circuit is *stabilized*.

The other cause of output voltage variations is change in the voltage applied across the input terminals of the circuit. If the input voltage should increase, the zener diode will pass a larger current so that the extra voltage is dropped across R_s. Conversely, if the supply voltage falls, the diode takes a smaller current and the voltage dropped across R_s is reduced. Because of the varying voltage drop across R_s, the load voltage fluctuates to a much lesser extent than does the input voltage.

Calculation of Series Resistance

(a) Varying load; Fixed supply voltage

When the load current varies, the current taken by the diode varies by the same percentage in the opposite direction. The diode current reaches its maximum value when the load current is zero, and at this point care must be taken to ensure that the maximum power dissipation rating of the diode is not exceeded. Therefore

$$I_{d(max)} = \frac{\text{Maximum power dissipation}}{\text{Diode (output) voltage}} \qquad (9.3)$$

The required value of R_s can now be calculated using equation (9.2).

EXAMPLE 9.1

A zener diode stabilizing circuit is to provide a 24 V stabilized supply to a variable load. The input voltage is 30 V and a 24 V, 400 mW zener diode is to be used. Calculate (i) the series resistance R_s required and (ii) the diode current when the load resistance is 2000 Ω.

Solution

(i) From equation (9.3),

$$I_{d(max)} = 0.4/24 = 16.67 \text{ mA}$$

From equation (9.2),

$$R_s = \frac{30-24}{16.67 \times 10^{-3}} = 360 \ \Omega \qquad (Ans.)$$

(ii) When the load resistance is 2000 Ω the load current will be $24/2000 = 12$ mA. The total current in R is 16.67 mA and

$$\text{Diode current} = 4.67 \text{ mA} \qquad (Ans.)$$

(b) Varying supply voltage; Fixed load

If the supply voltage to the stabilizer circuit should decrease, the diode current will fall so that a smaller voltage drop across the series resistor occurs. The zener diode must pass a minimum current $I_{d(min)}$ if it is to operate in its breakdown region and act as a voltage stabilizer. Therefore

$$R_s = \frac{V_{IN(min)} - V_{OUT}}{I_{d(min)} + I_L} \qquad (9.4)$$

EXAMPLE 9.2

A 9.1 V, 1.3 W zener diode has a minimum current requirement of 20 mA and is to be used in a stabilizer circuit. The supply voltage is 20 V ± 10% and the constant load current is 30 mA. Calculate (i) the series resistance required and (ii) the power dissipated in the diode when the supply voltage is 22 V.

Solution
(i) From equation (9.4)

$$R_s = \frac{18 - 9.1}{(20 + 30) \times 10^{-3}} = 178 \; \Omega \qquad (Ans.)$$

(ii) When $V_{IN} = 22$ V

$$I_d + I_L = \frac{22 - 9.1}{178} = 72.47 \; \text{mA}$$

$$I_d = 72.47 - 30 = 42.47 \; \text{mA} \qquad (Ans.)$$

and therefore

$$\text{Power dissipated} = 9.1 \times 42.47 \times 10^{-3} = 387 \; \text{mW} \qquad (Ans.)$$

(c) Varying supply voltage; Varying load

The maximum value of the series resistor R_s is determined by the necessary minimum diode current, and its minimum value is calculated to be large enough to ensure that the rated power dissipation of the diode is not exceeded. The value chosen for R_s must be a compromise between the minimum and maximum values.

Conversion of D.C. Supplies

Many equipments, particularly portable ones, are required to operate from a low-voltage d.c. supply and there is a need for d.c./a.c. invertors and d.c./d.c. convertors. A d.c./a.c. INVERTOR is a circuit that converts a d.c. source into a required a.c. supply and a d.c./d.c. convertor is a circuit that converts a d.c. source at one voltage into a d.c. supply at another voltage. A d.c./d.c. CONVERTOR consists of a d.c./a.c. invertor followed by a suitable rectifier as shown by the block schematic of Fig. 9.16.

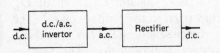

Fig. 9.16 Principle of d.c./d.c. conversion

Transistor D.C./A.C. Invertors and D.C./D.C. Convertors

Two types of d.c./a.c. invertor are in common use: firstly, there is the ringing-choke convertor and, secondly, there is the push-pull convertor. The push-pull has the advantages of requiring a smaller and cheaper transformer and of giving a higher output power with better voltage regulation, but it has the disadvantages of lower efficiency and of requiring two transistors as opposed to only one.

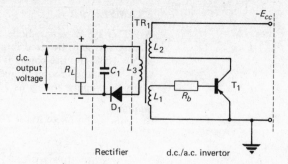

Fig. 9.17 The ringing-choke d.c./
d.c. convertor

The basic circuit of a RINGING-CHOKE d.c./d.c. convertor is shown in Fig. 9.17, and consists of a push-pull d.c./a.c. invertor followed by a half-wave rectifier. When the circuit is first switched on, the transistor is biased so that the resistance between its collector and emitter terminals is low. Practically the whole of the supply voltage is then applied across winding L_2 and a current flows in this winding that increases linearly with increase in time. A constant voltage is induced in winding L_1 and so a constant base current is provided for the transistor. The magnitude of this current is determined by the value of the resistor R_b. Diode D_1 is non-conducting during this period. The increase in base current changes the bias (operating) point of the transistor, and immediately this point coincides with the knee of the output characteristic the collector/emitter resistance increases abruptly, and the current flowing in winding L_2 falls. The sudden decrease in the current flowing in L_2 reverses the e.m.f. induced in winding L_1 and this causes the base current to fall rapidly. The transistor then rapidly ceases to conduct, or cuts off. The rapid change in the current flowing in L_2 induces a large voltage in winding L_3 and this voltage rises rapidly, charging capacitor C_1 via diode D_1, until the capacitor voltage has risen to the output voltage. The energy that was stored in the magnetic field has now been transferred to the dielectric of the capacitor. The voltage across L_2 now collapses (reverses direction) and induces a voltage in winding L_1 that switches T_1 on again and restarts the cycle. The d.c. output voltage is the voltage appearing across capacitor C_1 and it can be smoothed if necessary.

When a high output voltage is required, a very high turns ratio is necessary for the transformer and this leads to difficulties in transformer design. The use of a voltage-multiplying circuit is then advantageous.

Ringing-choke convertors are used for low output powers and mainly for d.c. outputs.

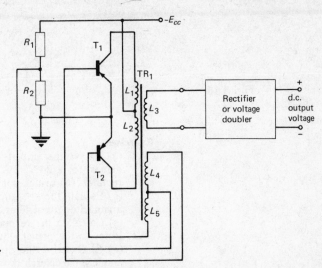

Fig. 9.18 The push-pull d.c./d.c. convertor

Fig. 9.18 shows the basic circuit of a PUSH-PULL d.c./d.c. convertor; the block marked "rectifier" may consist of any form of full-wave rectifier or voltage multiplier. When the supply is switched on, the potential divider, $R_1 + R_2$, provides a base current for transistors T_1 and T_2 that biases them into a state of conduction. Because of slight unbalances in the circuit, and in the transistors themselves, one transistor conducts more heavily than the other, and this results in the circuit rapidly going into the condition of one transistor fully conducting and the other completely non-conducting. Imagine transistor T_1 to be fully conducting and T_2 to be non-conducting. The collector-to-emitter resistance of T_1 is then small and so practically the whole of the supply voltage appears across winding L_1. This causes the current in L_1 to rise linearly until the point is reached where the transformer core is magnetically saturated. The current then rises suddenly and rapidly to try to maintain the same rate of change of flux in the core, and a large voltage is induced in winding L_3. The rise in current is limited by the circuit parameters and, when this limit has been reached, the rate of change of flux can no longer be maintained and the flux decays to zero. The voltages appearing across the windings collapse and T_1 is switched to the non-conducting state. The current then ceases to flow in winding L_1 and the resultant decrease in core flux induces voltages of opposite polarity in the other windings. Transistor T_2 is then switched to the fully conducting state and the cycle repeats in the opposite direction. The alternating voltage that appears across winding L_3 may be either fed directly to the load or applied to a rectifier as shown. If necessary, voltage multiplying and/or smoothing may be used.

Push-pull convertors are normally operated at a frequency of some 300–2000 Hz and can produce output voltages of up to several kilovolts from a few volts input.

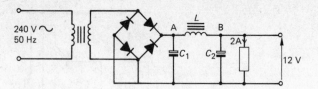

Fig. 9.19

Exercises

9.1. (a) Draw the circuit diagram, including typical component values, for a 24 V, 2 A power supply based on (i) a half-wave rectifier, (ii) bridge-connected rectifiers.

(b) Explain the operation of and draw a regulation curve for (i) a capacitor input filter, (ii) a choke-input filter.

(c) Calculate the reactance of a 10 μF capacitor at 50 Hz.

9.2. Refer to the power supply shown in Fig. 9.19.

(a) (i) Describe the operation of the circuit comprising C_1, L and C_2. (ii) Sketch the voltage waveforms present at points A and B.

(b) Describe how an output voltage regulation curve may be obtained for a power supply. Detail the equipment required and sketch a typical regulation curve indicating the units on each axis.

(c) Calculate the reactance of 10 H at 100 Hz. (C & G)

9.3. (a) Draw the circuit diagram of a full-wave power supply to provide 0.5 A at 20 V from 240 V, 50 Hz mains. List separately a suitable value for each component, including the transformer.

(b) State the frequency of the ripple which may be present and explain how this arises.

(c) Describe how the output regulation of a power supply may be obtained and sketch the result expected with the values indicated. (C & G)

9.4. A zener diode stabilizing circuit operated with an input voltage of 20 V, a diode current of 10 mA to provide 10 V across a 1500 Ω load. Calculate (i) the value of the series resistor and (ii) the diode current when the input voltage rises to 22 V.

9.5. A zener diode stabilizer provides 18 V output from a 22 V supply. The load resistance is 2500 Ω and the current flowing in the zener diode is 12 mA. Calculate (i) the value of the series resistor, and (ii) the diode current when the load resistance is reduced to 2000 Ω.

9.6. Draw the circuit of a half-wave rectifier with a capacitor-input filter. Give typical component values and describe the operation of the circuit. What is the peak inverse voltage of the diode (i) without the capacitor connected and (ii) with the capacitor connected?

9.7. What is meant by the *regulation of a power supply*? Give an expression for the percentage regulation. What constitutes good regulation and why is it desirable?

The output voltage of a power supply when no current is taken by the load is 24 V. When full load current is taken the output voltage falls to 22 V. Calculate the percentage regulation of the power supply.

9.8. Draw the circuit diagram of a choke-input filter for a power supply and describe its action. Such a filter has an inductance of 10 H and a capacitance of 10 μF. Calculate the percentage reduction in ripple voltage the filter provides if the ripple frequency is (i) 50 Hz and (ii) 100 Hz.

9.9. A zener diode stabilizer circuit provides 30 V to a load that may vary from 1000 Ω to 500 Ω. The supply voltage is 50 V and a 30 V, 1.3 W zener diode is used. Calculate (a) the series resistance required and (b) the diode current when the load resistance is 3000 Ω.

9.10. A 10 V, 1.5 W zener diode is used in a voltage stabilizing circuit. The input voltage may vary between the limits of 20 V to 22 V and the load is constant at 20 mA. Calculate (a) the series resistance required if the diode current must not fall below 4 mA and (b) the power dissipated in the diode when the input voltage is 21 V.

Short Exercises

9.11. Sketch waveforms of applied a.c. voltage and load current for diode circuits which provide half-wave and full-wave rectification into a resistive load.

9.12. Draw the output voltage waveform of a half-wave rectifier and then show the effect on this waveform of connecting a capacitor across the load resistance.

9.13. Repeat exercise 9.12 for a full-wave rectifier circuit.

9.14. Answer the following questions on a half-wave rectifier circuit.
(a) What is the peak inverse voltage?
(b) What is the effect on the output waveform of (i) increasing the reservoir capacitance and (ii) decreasing the load current?

9.15. List the advantages of the bridge rectifier circuit over the half-wave rectifier circuit.

9.16. Draw the regulation curves for capacitor and choke-input filters. Explain why choke-input filters often use a *swinging choke*.

9.17. Draw the circuit diagram of a zener diode stabilizer and explain how it works (i) when the input voltage is constant and the load current varies, and (ii) when the load current is constant and the input voltage varies.

9.18. The peak voltage across the secondary winding of a half-wave rectifier input transformer is 20 V. (a) What is the approximate value of the output voltage when no current is supplied to the load; (b) What is the peak inverse voltage?

9.19. A power supply unit has a no-load output voltage of 9 V and a percentage regulation of 1%. Calculate the full load current into a load of 500 Ω.

9.20. Explain the meanings of the following terms used in connection with electronic power supplies: (a) regulation, (b) peak inverse voltage, and (c) reservoir capacitance. List the advantages and disadvantages of using a battery to power a small electronic equipment.

Numerical Answers to Exercises

1.3 1865 Ω-m, 1213 Ω-m

2.7 18 μA

2.8 133.3 mA

3.12 $A_v = 98$, $A_p = 4802$

3.14 $h_{fb} = 0.9934$, $h_{fc} = 151$

3.22 $A_v = 183.3$, $A_p = 18\,333$

3.23 $I_{CEO} = 1.5$ mA

5.7 -640 V, OV, $+300$ V, $+600$ V, $R_1 = 260$ kΩ, $R_2 = 40$ kΩ, $R_3 = 600$ kΩ, $R_4 = R_5 = R_6 = 300$ kΩ

5.15 41.67 V/cm **5.18** 75 V

6.13 12.25 V **6.17** 163.6

7.1 105 pF to 422 pF **7.2** 4

7.3 18.8 pF to 56.3 pF **7.4** 1.1 MHz

7.5 15.6 pF to 62.5 pF **7.7** +14.1%

7.8 43 pF, 678 pF **7.9** 0.67, 0.4

7.12 7.958 kHz, 15.915 kHz **7.17** 1 MHz

7.21 1, 0.5

8.5 $A = 0$ or 1, $A = 1$ **8.19** $A = 0$

9.1 318.3 Ω **9.2** 6283.2 Ω

9.4 600 Ω, 13.33 mA **9.5** 208 Ω, 10.2 mA

9.7 8.33% **9.8** 88.7%, 97.4%

9.9 405.4 Ω, 19.3 mA

9.10 416.7 Ω, 64 mW **9.19** 17.82 mA

Electronics II: Learning Objectives (TEC)

(A) Elementary Theory of Semiconductors

(1) *Understands the simple concept of semiconductors.*

(2) *Knows behaviour of a p-n junction with forward or reverse bias.*

page 37 6.2 Describes the method of obtaining the common-base mode static characteristics.

  6.3 Discusses given typical families of curves of

39 I_c/V_{cb} (output characteristics)

39 I_c/I_e (transfer characteristics)

37 V_{eb}/I_e (input characteristics)

37 6.4 Sketches a common emitter mode test circuit-diagram for determining the static characteristics.

40 6.5 Measures the common emitter static characteristics.

  6.6 Plots and describes typical families of curves of

43 I_c/V_{ce} (output characteristics)

42 I_c/I_b (transfer characteristics)

41 V_{be}/I_b (input characteristics)

38, 40, 42, 43 6.7 Determines the values of h_{fb} and h_{fe} from given characteristics.

38, 41 6.8 Determines the value of input resistance from given input characteristics.

39, 42 6.9 Determines the value of output resistance from given output characteristics.

(B) Cathode ray tube

(11) *Knows the principles of operation of a cathode ray tube.*

71 11.1 Labels a diagram of a C.R.T.

  11.2 Explains the functions of the following:

71 (*a*) electron gun

72 (*b*) focus control

72 (*c*) intensity control

77 (*d*) blanking pulses.

73, 75 11.3 States that deflection can be produced by electric and/or magnetic fields.

77, 78 11.4 Demonstrates the use of timebases and of vertical and horizontal deflection controls.

(C) Small-signal amplifiers

(12) *Knows the circuit and operation of a small-signal common-emitter amplifier.*

85, 86 12.1 Draws the circuit-diagram of a single-stage amplifier having a load resistor R_L.

88 12.2 Shows that the supply voltage $V_{cc} = I_c R_L + V_{ce}$.

82, 85   12.3   Explains that bias is required to give a selected quiescent operating point on the output characteristic.

85   12.4   Sketches the circuit-diagram and explains the action of a simple bias arrangement consisting of a resistor connected between V_{cc} and base.

31, 33, 82, 89   12.5   Explains the effect of a small sinusoidal current input on the quiescent condition.

33, 82   12.6   States that voltage phase inversion occurs between input and output signals.

   (13)   *Constructs and uses a d.c. load line on transistor characteristics*

page 88   13.1   Constructs the load line on a given set of output characteristics of a common emitter amplifier for a stated value of load resistance.

89   13.2   Estimates the r.m.s. voltage output from the load line for given quiescent conditions and given input signal.

34, 89   13.3   Determines the voltage gain A_v from the static characteristics assuming a given input resistance.

89   13.4   Determines the current gain A_i from the static characteristics.

34   13.5   Calculates the power gain A_p in dB.

46   13.6   Describes thermal runaway of a transistor.

47   13.7   States the reasons for use of heat sinks.

   (16)   *Understands automatic biasing of small signal amplifiers.*

84   16.1   Explains, with the aid of sketches, simple methods of biasing a transistor amplifier stage.

(D) Waveform Generators

   (17)   *Knows typical oscillator waveforms.*

77, 96, 97   17.1   Sketches output waveforms of oscillators in common use: sinusoidal, rectangular, saw tooth.

page 97, 98   17.2   States the common uses of the waveforms set out in 17.1.

   (18)   *Knows the principles of simple sinusoidal oscillators.*

98   18.1   States that a sinewave oscillator is an amplifier with positive feedback sufficient to maintain its own output.

99   18.2   States that a sinewave oscillator requires both a frequency-determining circuit and a method of self-stabilization.

101   18.3   States that the approximate frequency of oscillation of most L-C sinewave oscillators is $f_0 = \dfrac{1}{2\pi\sqrt{LC}}$

103, 104 18.4 Sketches the circuit diagram of a tuned-collector oscillator.

101, 102 18.5 Describes methods of applying bias in a tuned-circuit transistor oscillator.

(F) Logic Elements and Circuits

(19) *Knows that information can be communicated by two-state signals.*

109, 110, 114, 115 19.1 Gives simple examples of two-state devices.
109, 110, 111, 112 19.2 Gives simple examples of information being communicated by two-state devices.

(20) *Understands the function of, AND, OR and NOT gates.*

116 20.1 States the logical function of the AND-gate.
117 20.2 Constructs a truth table for a 3-input AND-gate.
116 20.3 States the Boolean symbol for AND.
116 20.4 Draws the B.S. circuit symbol for an AND-gate.
116 20.5 Recognises superseded B.S. circuit symbols for an AND-gate
117 20.6 States the logical function of the OR-gate.
118 20.7 Constructs a truth table for a 3 input OR-gate.
117 20.8 States the Boolean symbol for OR.
116 20.9 Draws the B.S. circuit symbol for an OR-gate.
116 20.10 Recognises superseded B.S. circuit symbols for an OR-gate.
119 20.11 States the logical function of the NOT-gate.
119 20.12 Constructs a truth table for a NOT-gate.
119 20.13 States the Boolean symbol for NOT.
116 20.14 Draws the B.S. circuit symbol for a NOT-gate.
116 20.15 Recognises superseded B.S. circuit symbols for a NOT-gate.

(21) *Understands the action of simple electronic gates.*

page 121, 122 21.1 Explains the action of a 3-input diode AND-gate.
121, 122 21.2 Explains the action of a 3-input diode OR-gate.
114 21.3 Explains the action of a transistor when used as a switch.
115 21.4 Explains the action of a transistor NOT-gate.

Index